Alexander BOTTS

and the EARTHWORM TRACTOR

BOTTS ABROAD

William Hazlett Upson

Octane Press, Edition 1.0, April 2021
Volume 2

Main cover illustration by Nick Harris (Beehive Illustration)
Cover portrait by Walter Skor
Interior illustrations by Tony Sarg
Antique Tractor Graphic from Wisconsin Historical Society,
WHS-97419 (image altered)

 The stories in this series of books were originally published in *The Saturday Evening Post*. While these stories have been reprinted in other books, this book is the only one to have the stories in their entirety as well as the original artwork.

ISBN: 978-1-64234-039-6
epub ISBN: 978-1-64234-044-0
LCCN: 2021930668

Project Edited by Catherine Mandel
Design by Tom Heffron
Copyedited by Maria Edwards
Proofread by Dana Henricks

Austin, TX
octanepress.com

Printed in the United States

CONTENTS

FOREWORD

A Lifetime with Alexander Botts

BY CATHERINE MANDEL

Alexander Botts has been charming readers for nearly one hundred years. The world's greatest tractor salesman was imagined by William Hazlett Upson and appeared in a series of over one hundred short stories in The Saturday Evening Post *from 1927 to 1975. Mr. Botts was born out of the author's employment at the Caterpillar Tractor Company. He worked on the motor assembly line before transferring to demonstrations and sales. It was here that he encountered salesmen and agricultural workers who inspired the timeless, good-natured stories. In each tale, Mr. Botts goes to great lengths to make a sale and, more often than not, his plan goes awry. He pulls through in the end though, which is what keeps readers coming back for more.*

IN 1947, A TWELVE-YEAR-OLD in a barber shop in northern New Jersey picked up a copy of *The Saturday Evening Post*. Thumbing through, he stopped at the latest Alexander Botts story and read while he waited for a trim. The young Donald Stiner—who was already driving tractors on his family's farm—was immediately captivated by the "questionable strategies and abundance of good luck" of the world's best tractor salesman.

For the next twenty-five years, each time Mr. Stiner came across the weekly periodical, he flipped through looking for the newest story in the series. When the stories were presented in book form, he devoured those. By 2021, the eighty-five-year-old Mr. Stiner had amassed a nearly complete collection of the published works by Mr. Upson, including most bound collections of Alexander Botts stories.

Mr. Stiner describes the stories as intoxicating, and he is not alone in his fascination with Alexander Botts. The stories became a worldwide phenomenon, spawning book collections and movies. Mr. Upson combines wholesome, charming characters with uncomplicated, pleasurable storylines. The results are pure entertainment.

Mr. Stiner enjoys Mr. Botts's ambition and cleverness. Never mind the fact that each predicament Mr. Botts finds himself in is purely

self-inflicted, his inventiveness and resourcefulness are traits less utilized in the time of cell phones and the internet.

Another attractive story element is Mr. Botts's ability to tinker with his machines. Whether he knows what he's doing or not, he doesn't hesitate to get his hands dirty and dig into the guts of his crawler tractors. For Mr. Stiner and a legion of present-day readers, Mr. Botts's ingenuity and mechanical nature are part of the allure of the stories.

In 1970, Mr. Stiner traveled for work to Middlebury, Vermont, Mr. Upson's hometown. The Vermont Book Shop stocked Mr. Upson's story collections and even published a volume locally.

Mr. Stiner inquired about the author and his works and was shocked to hear the shopkeeper ask if he wanted to meet Mr. Upson. Much like the character from one of his stories, after a short phone call Mr. Upson invited Mr. Stiner to his home on a small farm with a John Deere tractor. They talked for more than an hour, and he left with two more signed books.

In a fictional biography of Alexander Botts, Upson writes of his hero: "He sincerely wants to help the customer. He never puts across a sale unless

"These horses started up. The panic spread. And before the unfortunate drivers knew what was going on, the entire thirty animals were galloping madly across the field, dragging the great harvester after them." (from the story "Horse Play")

the customer will benefit. He has courage, resourcefulness. He never holds a grudge. He is, in short, a good egg." After having read nearly everything written by Mr. Upson and meeting him in person, Mr. Stiner is inclined to say the same applies to the author.

It is refreshing to learn Mr. Upson was as accessible as his stories and the characters in them. He weaves imaginative narratives with effortless style to appeal to a wide audience. Part of Mr. Upson's charm is his sense of humor about the art of writing. In *How to Be Rich Like Me*, he facetiously suggests you simply write for radio.

He jokes about his lackadaisical habits, yet he's penned more than one hundred Alexander Botts stories alone, each unique and carefully written. Mr. Botts may be a bumbling salesman, but he never does anything half-heartedly and attacks every challenge with limitless enthusiasm. As much as he tried to pretend otherwise, Mr. Upson approached his craft of writing with an equal level of hard work and intensity.

To honor the timeless storytelling of William Hazlett Upson, Octane Press has acquired the full Alexander Botts series, some of which did not appear in The Saturday Evening Post. *The collection will be presented in its entirety for the first time along with the original illustrations.*

EUROPE IS DUE FOR A SURPRISE

ILLUSTRATED BY TONY SARG

EARTHWORM TRACTOR COMPANY
EARTHWORM CITY, ILLINOIS
OFFICE OF THE SALES MANAGER

JANUARY 19, 1928.

MR. ALEXANDER BOTTS,
DEANE SUPPLY COMPANY,
MERCEDILLO, CALIFORNIA.

DEAR BOTTS: At a meeting of the officers of the company this morning, it was decided to ask you to go to Europe for several months as our sales representative.

Since the merger last spring of the Farmers' Friend Tractor Company and the Steel Elephant Tractor Company into the new Earthworm Tractor Company, we have, as you know, greatly increased our business. And, as ample capital is available, we are planning an even greater expansion.

Up to this time our European business has been practically nothing. Economic conditions since the war have made it almost impossible for us to sell any tractors, either in England or on the Continent. But conditions have recently improved so much that we are seriously thinking of going into the European market.

Our plan is to send you and another man over at once to see what you can do. Your trip will be more or less experimental; possibly you may not be able to accomplish anything. But if you succeed in selling a reasonable number of tractors, we will know that the market is there, and we will open a number of European branches. If everything goes well, it is even possible that within a few years we may start a European factory.

We have arranged to send Mr. George McGinnis, whom you will remember as the former star salesman of the Steel Elephant Company, to England, Germany and the rest of Northern Europe. We want you to see what you can do in Southern Europe, particularly France and Italy.

We have not forgotten your splendid record when you were a salesman with this company, and we have noted with great satisfaction your fine work as tractor sales manager for the Deane Supply Company at Mercedillo.

Please let us know your decision at once. Your salary will be five hundred dollars a month plus expenses.

Cordially yours,
GILBERT HENDERSON,
Sales Manager.

DEANE SUPPLY COMPANY
MERCEDILLO, CALIFORNIA

JANUARY 23, 1928.

MR. GILBERT HENDERSON,
EARTHWORM TRACTOR COMPANY,
EARTHWORM CITY, ILLINOIS.

DEAR HENDERSON: Well, well, it seems like old times to get a letter from you saying that you want me to go out on a trip. The idea appeals to me very much. I have a good assistant here who can handle my job while I am gone, so there is no reason why I can't go. Of course, five hundred a month is nothing at all as compared with what I get from my share in the business here, but I realize that you can't afford to pay big money on what you regard as a speculative, experimental trip. Of course, I always look at these vulgar money matters in a very large way. I suppose I am something like Mr. Andrew Mellon in this respect—always ready to take on an interesting, worthwhile job, either as a cabinet member or as a tractor sales ambassador, regardless of the financial sacrifice involved. However, as soon as I start making sales in a big way over in Europe, I will expect you to pay me adequately.

I was rather amused at your statement that I might possibly make no sales at all. Of course, it is perfectly possible—in fact, probable—that this man McGinnis may make no sales in England and Germany. But you don't have to worry about me. There must be people in Europe who can either beg, borrow or steal the price of a tractor; if so, I will sell them. And it is with a feeling of the greatest pleasure that I look forward to this experience.

Selling tractors around this place has become too easy. The people of California are so intelligent, and are so well informed regarding tractors, that they just come in and buy them of their own free will. I have had so few difficulties for the past year and a half that I am getting soft and out of condition mentally. If there is as much sales resistance in Europe as you indicate, it will be just what I need to tone me up. Consequently, I will accept your offer on two conditions:

1. Mrs. Botts is to go with me, and the expense account will cover the expenses of both of us. It is absolutely essential that Gadget go along. (Note: My wife's real name is Mildred, but I always call her Gadget, because she is the most valuable and useful companion I have ever found.) She has a fine education and a wonderful brain, and since our marriage I have come to rely absolutely on her excellent business judgment. The combination of my energy and resourcefulness with her brilliant intellect makes a team that cannot be beat. Furthermore, she can act as interpreter. In spite of the fact that I am familiar with France, having spent more than a year over there in the A.E.F., I am forced to admit that my knowledge of the French language is a trifle weak. Gadget has never been to Europe, but she has studied at a splendid summer school in Vermont and she can talk it better than most of the French people themselves. She also knows Italian and German. Consequently, she must go along.

2. There must be no hollering, yawping, or nagging about small sums on my expense account. I am very insistent about this. The greatest annoyance connected with my otherwise pleasant association of former years with your tractor company was the continual bickering about my expense account. If I am to be at my best on this foreign trip, I must not be hampered and constricted in my style by petty money considerations. I am willing to be broad-minded. It is perfectly reasonable that you should check up closely on expense items running to a hundred dollars or more, but all trifling amounts, such as ten, twenty or fifty dollars, must be passed without question, and in a gentlemanly way.

If you agree to these two conditions, I will accept your offer with the greatest joy. When do we start?

Sincerely,
ALEXANDER BOTTS.

EARTHWORM TRACTOR COMPANY
EARTHWORM CITY, ILLINOIS
OFFICE OF THE SALES MANAGER

JANUARY 27, 1928.

MR. ALEXANDER BOTTS,
DEANE SUPPLY COMPANY,
MERCEDILLO, CALIFORNIA.

DEAR BOTTS: Your letter has come, and we are very glad to welcome you back as an Earthworm tractor salesman. We accept your two conditions. We have engaged passage for you and Mrs. Botts on the steamship *Beaucaire*, sailing from New York on February eighteenth, and due in Marseilles on March third. We are shipping on this same boat eight demonstration tractors—four of them billed to Marseilles to be used for your work in France, and four of them billed to Genoa—at which port the *Beaucaire* also touches—for your work in Italy. We are also including a combined harvester and an assortment of plows, blade graders, wheel scrapers, dump wagons and other machinery.

We are mailing foreign advertising matter to chambers of commerce and agricultural associations all over France and Italy. We will supply you with additional advertising matter, and with a number of reels of talking motion pictures showing tractors in action, with running sales talks in French and in Italian.

We are giving you the greatest latitude on this trip. We expect you to make inquiries and find out about conditions when you get there. We want you to sell tractors at any place, or in any way that you can, with a view to opening up the South European market for future sales. You will keep in touch with me at all times, and I shall expect full reports of your progress.

Very sincerely,
GILBERT HENDERSON,
Sales Manager.

ALEXANDER BOTTS
EUROPEAN REPRESENTATIVE
FOR THE
EARTHWORM TRACTOR

ON BOARD S.S. *BEAUCAIRE*.
FEBRUARY 20, 1928.

MR. GILBERT HENDERSON,
EARTHWORM TRACTOR COMPANY,
EARTHWORM CITY, ILLINOIS.

DEAR HENDERSON: We are now two days out from New York. So far it has been pretty rough, but Gadget and I are feeling fine, and already I have much to report. You don't know how glad I am that I am taking this trip. Already I am all steamed up and working hard. An ordinary salesman, on going to Europe, would probably wait until he got there before he tried to do any business. But I started in as soon as I got to New York—two days before the boat sailed. First of all, I had this letter paper made up. If you run your finger over the printing you will see that the letters stand out in a very expensive way. Furthermore, the paper is the highest-grade bond, in every way worthy of the European representative of such a high-grade machine as the Earthworm tractor. I have also had some high-grade business cards printed.

All this, of course, is of no great importance. I merely mentioned it so that you could see that I am right up on my toes, and attending to all details—even the most trifling—with my old-time vigor and efficiency.

While attending to the smaller matters, I have not neglected the larger, and you will be delighted to learn that I have evolved and commenced to put into action an idea that is as large and splendid as anything I ever put across in the good old days.

This idea came to me when I first reached New York. At the steamship office I learned that on this boat, the *Beaucaire*, there would be a delegation of forty or fifty French grape growers, who were returning to France after a visit to the United States. The steamship officials did not know the exact purpose of their visit to America, but it at once occurred to me that they had probably come to investigate modern American methods of vineyard culture. And if they had visited the splendid vineyards in California they would, of course, have seen hundreds of our small Earthworm tractors at work plowing and cultivating. They would have seen how far superior these machines are to any other method of culture. And if they had talked to the California vineyard owners, they would have learned that Earthworm tractors are the world's cheapest, easiest, and best method of taking care of grapevines. I saw at once that these French grape growers were my meat. I had a feeling that when I approached them I would find them already about half sold on the idea of Earthworm tractors.

An ordinary salesman, when he feels that a prospect is already half sold, usually slackens his efforts, hoping that he can swing the sale without much more work.

With me, however, it is just the opposite. When things begin to look favorable I always work twice as hard and exert every possible effort to make sure that nothing shall prevent matters from coming to a triumphant and favorable conclusion.

In this case I decided to catch these Frenchmen off their guard on the boat, while I had them when they couldn't get away, and put on a terrific high-powered selling campaign.

At once I rushed down to the pier, where the good ship *Beaucaire* was being loaded. I was delighted to find that the eight tractors which we are shipping had not yet been put on board. They were all standing on the pier, neatly boxed for export. I told the man in charge of the loading that only seven of these machines were to go in the hold; the eighth one—a small ten-horsepower model—I told him was to be placed on the promenade deck where it could be used for demonstration purposes.

The man seemed very much surprised and puzzled at this request. He said he had never heard of such a thing. So I started in and explained everything to him as patiently and politely as I could. I impressed it on his mind that this was a very reasonable request, that I would not hurt the ship in any way, that I would bring along a set of rubber pads to bolt onto the tracks so that the deck would not become scratched, that I would be very careful not to run over any of the passengers, and that most of these passengers were French vineyard men who would be deeply interested in the tractor and would want to see it working.

After I had elucidated this matter for more than half an hour, the man said he had no authority to load the tractors anywhere but in the hold, and that I would have to see one of the officials of the company.

Accordingly, I got him to promise to leave the small tractor on the pier for the time being, and want back to the company office. Here I found a Mr. Brown, who seemed to be in charge of the company's freight business.

I had to talk to him about an hour, explaining everything in great detail. Then he said I would have to see another guy whose name I have now forgotten. I talked and argued with this other guy for almost two hours, and then he told me that it was a matter which would have to be decided by the captain of the ship.

As the captain was French, I took Gadget along to act as interpreter—which was unnecessary, as it turned out that he spoke very good English. However, it was lucky that we both went, because the captain was very stupid and stubborn, and it took the best efforts of the two of us—talking alternately for about three hours—to convince him that we were right.

He finally agreed to put the tractor on the aft end of one of the upper decks, where it would be lashed down to prevent it from sliding around in rough weather. For two hours every afternoon—provided the weather was fair—he would have the end of the deck roped off and let us drive the tractor around and show what it could do. I had to sign a paper releasing the steamship company from all responsibility in case anything happened to the tractor, and assuming full responsibility for any damage the machine might do. He absolutely refused to give me the run of the whole deck with the machine. And for these restricted privileges I had to pay three hundred dollars—which I suppose is for deck rent or something. This seems like a pretty stiff price, but it is simply worthwhile, in view of the tremendous impression we are sure to make on these Frenchmen.

So far it has been too rough to do anything. Gadget and I are feeling great, but practically all of the Frenchmen seem to be cooped up in their staterooms. Apparently they are seasick—which, of course, is of no importance, because the boat is rolling so heavily that we could not demonstrate the tractor to them anyway. But I am all prepared to put on a swell demonstration as soon as we get some fair weather. I have rubber pads installed on the track shoes, and these ought to give me good traction on the boards of the deck. I have a number of railroad ties which I can pile up in various ways so that the tractor can climb over them and show these Frenchmen how it negotiates rough country. Also, I have fixed up a system of pulleys with large weights for the tractor to lift. This will give them an idea of what we can pull. Toward the end of the voyage I expect to put on a moving-picture show, showing the tractors doing actual farm work. The projector on the ship is equipped with the latest sound system, and that should help a lot. I am much encouraged. Never before, as far as I know, has there been a tractor demonstration on a great transatlantic liner. It is sure to make a tremendous impression. Already I can see a flood of orders pouring in.

February 21, 1928.
Three days out.

Still too rough to do anything with the tractor. But it is a little smoother than yesterday, and the Frenchmen are commencing to come up out of their holes. Some of them have their wives with them. Gadget and I have made the acquaintance of several couples. And Gadget, owing to her superb knowledge of the French language, has unearthed some information which indicates that our projected demonstration will be of even more importance than I had supposed. Apparently we shall have to start at the very beginning in our education of these people. They know nothing of American methods, and they care less. They say that Frenchmen have been growing grapes since before the Year One, and know all about it; and it would be absurd to think they could learn anything from California, where the large vineyards are less than a hundred years old.

They had not come to America to learn; they had come to instruct. They had been to Washington, and their mission was to persuade the American Government that the prohibition law was all wrong. In the first place, they said, it deprived them of one of their best markets, and thus seriously cut into their profits. In the second place, it deprived the American people of the world's most pleasing and healthful beverages.

To prove this better point they had brought along a full line of samples. They had supposed that the official nature of their visit, and the fact that they had a letter from the President of France, would let them take this stuff into the United States through some sort of diplomatic courtesy. But the customs authorities had been hard-boiled and had decided against them. And the wines are now on their way back to France on this very boat.

In spite of the lack of samples, however, they had pointed out to everyone they met that the French light wines are a natural God-given drink, completely harmless, an aid to digestion, a promoter of good fellowship and good feeling, and, when properly used, do not cause intoxication. They had, therefore, suggested—in the friendliest and most helpful spirit, and making it plain that they had no wish to meddle in our affairs—that it would be a great benefit to all concerned if the law were amended in such a way as to admit French wines to the United States. If this were done, prosperity would settle once more over the fair vineyards of France, and the American people, instead of being tempted to indulge in white mule and other vile distilled liquors, could return once more to the smooth and satisfying juice of the grape. They had spread this message among all the senators, representatives and officials they met, and everywhere they had been received with such politeness and good will that they felt sure—in spite of the fact that no one had promised them anything definite—that the laws would soon be changed in their favor.

Gadget and I, of course, had our doubts about this, but we were too polite to mention them to the Frenchmen. I did, however, discuss the matter with an American by the name of Bowers that I met in the smoking room.

"It seems incredible," I said, "but these grape growers actually think that all they have to do is tell Senator Heflin and the rest of them what wonderful stuff this French wine is, and right away they will drop everything else and push through a law so that it can be brought into the country. For pure, childlike simplicity you couldn't beat that anywhere."

"Well," said Bowers, "it doesn't seem to me anywhere near as simple-minded as your idea that you can go over to Europe, tell the Europeans what wonderful machines your Earthworms are and then sit back and watch the orders come rolling in. I doubt if you sell a single one."

"Why not?" I asked.

"The Europeans don't like machinery. They hate it—especially if it's American. I've traveled in Europe a lot and I know."

"You seem to be pretty sure of yourself," I said.

"I am," he said. And then he told me a long yarn about an American he knew that went broke trying to sell some kind of motor to the gondoliers in Venice. It seems this guy had it all figured out how he would speed up traffic in the Grand Canal several hundred percent, and he had thought out a system of traffic control with red and green lights on the Rialto Bridge and at other strategic points. But the Venetians couldn't use it at all; they passed a law prohibiting that kind of motor anywhere in the city, and the poor man didn't make a single sale.

"Probably," I said, "he was a poor salesman. But I am different. And if you want to see a selling campaign handled right, you just want to stick around when I put on my magnificent tractor demonstration on the promenade deck. If you want to see names being signed on dotted lines, that will be your big chance."

"I'll believe it when it happens," he said.

"Wait and see," I said.

February 22, 1928.

Washington's birthday, and a special dinner, with flags on the tables. The sea is getting smoother all the time. It won't be long now.

SUNDAY, MARCH 4, 1928, 8 A.M.

Since the last entry in this letter we have had a most distressing time. On the evening of Washington's birthday it started to get rough. The next day we were in the midst of a regular storm, which continued the rest of the way across the ocean, and even after we got into the Mediterranean. I had never supposed that a ship could heave and bang around the way this one did. The French passengers—and, in fact, almost everybody else—remained cooped up in their staterooms both day and night. And part of the time even Gadget and I felt a bit wobbly. It has been absolutely impossible for me to do any business at all.

But today I believe that my great hour is at hand. Yesterday afternoon it began to clear up a little, the wind died down, and this morning is as fair and beautiful a day as I have ever seen. There is still a little ground swell, but not enough to do any harm. This is the first and only day of the entire voyage that the sea has been smooth enough to make possible a tractor demonstration. The Frenchmen are all up on deck looking at the beautiful blue sky and the beautiful blue water, and enjoying the balmy, warm spring breezes. We are due to land at Marseilles late this afternoon—one day late.

I wanted to start demonstrating the tractor first thing this morning, but the captain would not permit it. He quoted our arrangement, which provided that I should drive the tractor for two hours in the afternoon only, on such days as the weather permitted. I told him that, in view of the fact that I had lost out on all afternoons except this one, I ought to be given a morning period, but he is a stubborn old cuss, and I couldn't make him see my point of view at all.

However, a short, snappy, intensive demonstration is often better than a series of dull, uninteresting and long-drawn-out ones, so I have the highest hopes of making a sensational impression. Gadget is already mingling with our charming French friends, and I will join her in a moment. The canvas cover has been taken off the tractor, and we will spend the rest of the morning letting these French people look at it and telling them its good points.

At noon the French delegation is giving a dinner for all the passengers. This dinner is dedicated to the honor and glory of French wines, and they are going to serve some of the excellent samples which they carried to America and which—fortunately for us—they were unable to get into the country. After the meal, I have arranged to show my talking moving picture of tractors in action, following which we will all assemble on deck, where I expect to astound them all by letting them see what a real tractor can do. Following the demonstration, Gadget and I will circulate through the crowd with order blanks so that the Frenchmen can put their names on the dotted lines. I have a feeling that you are going to be greatly pleased and surprised at the final installment of this letter, which I shall probably not have time to write until after we have landed.

MARSEILLES.
MONDAY MORNING, MARCH 5TH.

I have just been reading over the last few sentences which I wrote yesterday, and I hardly know how to begin my narrative of the events which took place on our last day aboard the good ship *Beaucaire*. You will, no doubt, be surprised when you hear what happened, but I fear you will not be as pleased as I had hoped you would be.

I am not pleased myself. As I sit here in this hotel room, weary and bruised in mind and body, my only pleasant thought is a sense of gratitude that dear old Gadget is here with me to cheer me up, and to change from time to time the bandage on my eye. This is the first time in my life that I ever had what might be called a real one hundred percent black eye, and I find that it is as painful as it is disfiguring.

The events of yesterday were most complicated. The day started most auspiciously, as I wrote you, with beautiful weather, a smooth sea, and everybody feeling fine. Gadget and I spent the morning talking tractors to dozens of Frenchmen. They were all very polite and very much interested, although perhaps a trifle noncommittal.

At noon the French delegation gave their dinner, which was graced by copious samples of the finest of wines. Various members of the delegation made speeches expressing their friendship and high regard for the American people, and their hope that these splendid beverages would soon have free entry into the greatest and most glorious of all the countries of the earth. There was a tremendous amount of applause for those sentiments, and everybody seemed to be in a remarkably pleasant and friendly frame of mind.

Unfortunately, however, these high-grade French wines are so delicate, and slide down the esophagus so smoothly and easily, that persons who are used to our more corrosive American bootleg stuff fail to realize that there is a considerable amount of kick concealed in these gentle liquids. The French people, of course, knew exactly what they were doing, but I fear that some of our American friends may, perhaps, have been a little unwise.

As for myself, I drank only water, as I wanted a clear head and a steady hand for the demonstration which was to follow. After the banquet was concluded, I arose—according to the program which had been arranged—and announced that the curtains would be drawn over the portholes and the room darkened so that we could have the talking motion pictures of the great Earthworm tractor. I made a few remarks to the Frenchmen who could understand English, requesting them to translate the gist of what I had to say to those who were so unfortunate as to know only French. I told them what a wonderful machine I had, how admirably adapted it was to vineyard culture and I laid great stress on its simplicity of operation.

"After the pictures," I said, "we will adjourn to the deck and I will show you what the tractor can do. And to prove to you how simple it is to operate, I will permit you to drive it around yourselves, if you so desire."

At this point, one of the American passengers arose and started to speak. He was a very large man with a florid face and red hair. I have since learned that his name is Mr. Tilton. He had been seasick and had stayed in his room practically all the way across, so I had never seen him before. At this time, however, he seemed to be full of life.

"I want to drive the tractor," he announced in a loud voice.

"Very well," I said. "Nothing would give me greater pleasure. As soon as the demonstration is under way, you may take your turn with these other gentlemen."

"I want to drive the tractor," said Mr. Tilton. "I want to drive it right away."

I noticed that the gentleman was swaying about a bit more than seemed justified by the very gentle motion of the boat, and I began to suspect that probably he had been indulging himself not wisely, but too well. Subsequent events showed that I was right in this surmise.

"I want to drive the tractor," he repeated. "I want to drive it right away."

"I am sorry," I said, "but we are going to see the moving pictures now. After that we will drive the tractor."

"I want to drive the tractor. I want to drive it right away. Listen," he went on, "I know all about those machines. I used to drive one of them when I was in the Army. And I just love those machines. They're wonderful—positively wonderful. It's ten long years since I've had a chance to drive one of them. And now that I have a chance, you won't let me do it. I am surprised at you."

"If you would only wait," I began, "until after the moving picture—"

"I want to drive the tractor," said Mr. Tilton. "I want to drive it right away."

"I'm surprised at you," he went on. "Here we are at this splendid banquet, given by our splendid French friends, where everything ought to be peace and good feeling and friendship—and what do you do? You come in and start an argument."

"But I didn't start an argument," I said.

"You're a liar," said Mr. Tilton. "You've been arguing with me here for five minutes. It's an outrage. I won't stand for it."

He brought his fist down on the table with a bang, and I noticed several stewards moving in his direction. As I did not want a painful scene, I walked over to him myself to try to quiet him.

"I want to drive the tractor," he said.

"Very well," I said. "Come with me." He followed me out of the dining saloon and up to the deck, where I prevailed upon him to sit down in a steamer chair. "You stay right here," I said, "and after the moving pictures I'll come back and we'll see what we can do about the tractor."

I then returned to the dining saloon and told the moving-picture operator to shoot. He did so, and I was immediately subjected to what I can only describe as a very painful surprise, caused by some very sloppy work at the New York office of the Earthworm Tractor Company. When I got those moving pictures at the New York office, they told me they showed the Earthworm tractor performing a great variety of tasks—agricultural and otherwise—with a running explanation and sales talk in the

French language. There were five reels, each in a large tin can, bearing the trademark of a well-known moving-picture company, and the label, "Earthworm Tractor Pictures. French." Naturally, I took these films in good faith, little dreaming that somewhere—probably at the motion-picture offices—someone had made a careless mistake and put in the wrong film.

You can imagine my astonishment when the picture opened with a loud burst of jazz music, and the title was flashed on the screen: "Red-Hot Tamales. A Rip Snorting Musical Spasm in Five Reels—All Talking, Dancing, Singing." Then, instead of an Earthworm tractor pulling a plow, there appeared a bevy of scantily clad young ladies stepping and kicking high, wide and handsome.

As fast as I could, I rushed around to the moving-picture booth and asked the operator what he thought he was doing. He protested that he was only running the pictures which I had given him, and this proved to be the case. Believe me, when I get back to the New York office I will tell those birds what I think of them. At once I stopped the show and explained to the people what had happened. Instead of being disappointed, however, they immediately set up a cry for the picture to go on. So, as I was most anxious not to offend anyone, I had to run the whole five reels. I think it was probably a pretty good show. The audience seemed to like it, but I was in no shape to enjoy the entertainment, and my annoyance was increased by Mr. Bowers, the pessimistic American, who sat down beside me and began telling me how all the American salesmen he had ever met were complete failures when it came to selling machinery to Europeans. He had a long sob story of a man that tried to sell electric refrigerators in Spain, and did no business at all. The Spaniards didn't even know what an old-fashioned ice refrigerator was. Another guy wore himself out trying to sell oil burners for furnaces to the Italians.

"And the Italians wouldn't even look at them," said Mr. Bowers. "They didn't even know what a furnace was, let alone a fancy oil burner."

"All of which," I said, "doesn't interest me in the slightest."

That shut him up for the time being. At last the moving picture came to an end, and I stood up and announced: "We will now go on deck for the big tractor demonstration."

As Gadget and I mounted the stairs, with the rest of the audience trooping along behind us, my heart beat high with hope. Although I was keenly disappointed that the moving pictures had contributed nothing to our tractor-selling campaign, I felt that we could make up for this lack as soon as we got the tractor in action. Little did I think, as I mounted the stairs, that the tractor was already in action.

My first intimation of disaster came when Gadget and I reached the door leading out onto the forward part of the promenade deck. This door was suddenly pulled open. There were loud shrieks of terror and dismay, and five or six elderly ladies came scrambling in. These ladies, we learned later, considered themselves rather old and feeble. They had not been feeling very well, so, instead of attending the banquet and moving-picture show, they had been resting in their steamer chairs. But when we saw them they were resting no more. As they came in that door they showed an amount of energy and agility that was positively remarkable.

I stuck my head out the door. I heard the roar of a motor. And looking toward the stern I was horrified to see my demonstration tractor approaching along the deck. In the driver's seat was Mr. Tilton, waving his hat with one hand and steering with the other. The gears were in high, the throttle was wide open and the machine was coming at top speed. Naturally, I was much displeased to observe Mr. Tilton taking such liberties with my tractor, and I was particularly annoyed at the course he was steering.

I might explain that the promenade deck was very wide. On the inside next to the cabin was a long row of about a hundred steamer chairs, all of them at this time vacant. Between these chairs and the rail was an open space, plenty wide enough for the passage of the tractor. But, instead of following this obvious path, Mr. Tilton had seen fit to veer over next to the cabin wall, and he was clattering along at a fearful speed, right down the middle of the row of chairs. If I live to be a hundred years old I will never forget that sight, with its fearful accompaniment of sound effect—the roar of the motor and the steady, sickening crash of rending wood, as one steamer chair after another was crushed under the whirling tracks of the machine.

As Mr. Tilton approached the door, I drew my head safely inside, and as soon as he had passed, I rushed out and pursued him. Ordinarily, I could have caught him very quickly, but in this case I kept stumbling and tripping over the fragments of the steamer chairs. In fact, I did not draw even with him until he had reached the forward end of the deck, swung around under the bridge and started down the deck on the other side of the boat.

"Stop!" I yelled, and reached out one hand to try to turn off the ignition.

"Aha!" said Mr. Tilton. "So it's you, you big bum!" And with a mighty swing of his large and powerful right arm he brought his fist around and

"I was horrified to see my demonstration tractor approaching along the deck."

landed what was by far the heaviest jolt I have ever received. It got me right on my poor unprotected left eye. And that was the last I saw of Mr. Tilton and the tractor.

Apparently I was knocked out cold. And I must have stayed out for some time. When I finally woke up, I was in my stateroom and good old Gadget was taking care of me. And as she insisted that I keep quiet, I did not learn what had happened until after we were inside the port at Marseilles, and most of the passengers had left the boat.

Finally Gadget consented to tell me that Mr. Tilton had made three complete circuits of the boat, wrecking every chair on the promenade deck, before they had been able to grab him and haul him off the machine.

After this it had been impossible to get any of the Frenchmen to listen to reason. Gadget had bravely rushed around and talked and pleaded with them, thrusting order blanks under their noses, and working upon them as hard as she could. But they were all in such a state of excitement that it was impossible for any mere human—even so remarkable a human as Gadget—to make any impression on them. All they would do was wander about, looking at the pitiful fragments of the steamer chairs, and commenting upon what a dangerous thing an Earthworm tractor was.

"Look what that terrible machine did," they said.

At this some of the American ladies came to Gadget's defense.

"It was not the fault of the machine," they said. "It was the fault of your cursed wine. The man was drunk."

Upon hearing this, one of the Frenchmen said, "That is an insult to the fair name of France, and to her honorable grape growers. If the man was drunk, it is the fault of the American prohibition law, which made it impossible for him to learn how to drink like a gentleman."

Then another Frenchman, for no reason at all, stood up and said, "You Americans are all crazy anyway, and you absolutely ruin the stewards by tipping them too much."

These statements started a long, bitter and completely idiotic argument between and among practically all the passengers on the boat. Gadget, of course, tried to quiet things down, as she is familiar with the well-known business principle that you never make any sales while engaging in a fight. But she had no luck. Everybody was too excited. And finally all the Frenchmen left the boat without placing a single order.

"But never mind, Alec," said Gadget. "We may be licked for the moment, but as soon as we get going on shore we will show these foreigners what kind of stuff we are made of, and we will get some real results."

"In the words of the immortal John Paul Jones," I said, "'we have only begun to fight!' And by the way, what happened to this guy Tilton?"

"That," said Gadget, "is the one bright spot in this whole mess."

"What do you mean?"

Gadget's answer shows that it is indeed a lucky thing I brought her along on this trip. "They took him down to his stateroom," she said, "and gave him a cold bath. He sobered up very quickly and came around and apologized to me most profusely. He really seems like a very nice man."

"When he is sober," I said.

"Yes. He had already settled with the captain for the damage he had done to the ship, and he offered to do anything he could for us to make up for the trouble he had caused. So I talked to him, and found out that he is a big contractor from Omaha. I asked him if he could use any Earthworm tractors in his business, and he said he could. So I sold him three machines. One of them is the tractor he was driving this afternoon—he had offered to pay for that anyway—and the other two are sixty-horsepower machines, to be delivered to him in Omaha next June. Of course, this sale doesn't count for anything on the foreign business that we are trying to get. But after all, it is something."

"It certainly is," I said. "But what is this guy going to do with the tractor over here?"

"What tractor?"

"Why, the one he was driving this afternoon. Didn't you say he just bought it?"

"Yes, he bought it," said Gadget. "And he paid for it. But there isn't much he can do with it. You see, when they grabbed him and hauled him off the seat of the tractor, they didn't succeed in getting the motor stopped. The machine ran right down the deck all by itself, broke through the rail and dropped into the water. So, in one way, I guess I have made something of a sales record. In all the history of the business I don't believe there is another case of anybody selling a tractor which was on the bottom of the Mediterranean Sea."

MONUMENT HISTORIQUE

ILLUSTRATED BY TONY SARG
OPENING ILLUSTRATION BY NICK HARRIS

ALEXANDER BOTTS
EUROPEAN REPRESENTATIVE
FOR THE
EARTHWORM TRACTOR

GRAND Hôtel DE L'UNIVERS, MARSEILLES.
MARCH 7, 1928.

MR. GILBERT HENDERSON,
EARTHWORM TRACTOR COMPANY,
EARTHWORM CITY, ILLINOIS.

DEAR HENDERSON: It gives me great pleasure to report that during the three days since we landed we have accomplished a tremendous amount of work; and Gadget and I have mapped out and already started to execute a comprehensive campaign by which we are going to make all of France tractor-conscious. (Note: I forget whether I have ever told you, Henderson, that "Gadget" is the nickname by which I usually refer to Mrs. Botts—her real name, Mildred, being too formal and nonmechanical for the tractor business.) I can assure you that you made no mistake in sending both Gadget and myself on this tractor-selling tour. Gadget has already more than paid for the expense of her passage by her assistance to me as interpreter and as general business adviser. Although the voyage over here was rather disagreeable, and although we failed in our attempts to sell any tractors to the various French passengers on the boat, we both have the brightest hopes for the future.

I spent all of Monday resting and recovering from the effects of the trip, and discussing plans with Gadget. We have decided that we will get much better publicity, and will provide a much finer educational treat for the French population, if we drive our demonstration tractors from place to place, instead of traveling by train. As long as our mission in this country is to instruct the French people in the advantages of using Earthworm tractors, it would be foolish for us to hide these machines inside the forty-*hommes*-eight-*chevaux* boxcars.

It is our plan to move north with the springtime, which has already touched this region with its verdure, but which has not yet, we understand, made itself felt to any great extent in Northern France. We will drive first to the city of Montpellier, which is a little less than two hundred kilometers from here, and which is the center of the largest grape-growing region in all France.

On the boat we met several big grape and wine men from Montpellier. We failed to make any impression on them during the voyage, but it is our intention to beard these babies in their own den. And after we get through subjecting them to a real high-pressure American sales campaign, they will have to be a whole lot tougher than they look if they escape ordering at least three or four machines.

We shall probably continue our assault upon the citizens of Montpellier and vicinity for about a week. Then we expect to swing back and advance up the valley of the Rhône, reaching Lyons early in April, and then proceeding on through Dijon, and reaching the famous wheat fields along the Marne sometime in May. After a brief but snappy selling campaign in that region, we shall probably hop on the train, rush down to Genoa, pick up the four demonstration tractors which have been shipped to that port and start a whirlwind tour through the valley of the Po.

Of course, these plans are merely tentative. We shall make constant side trips and short scouting expeditions. We shall pick up as much information as we can, and if conditions demand it, we shall not hesitate to change our itinerary.

Our general plan of campaign was complete Monday night. On Tuesday morning I felt completely rested up. And as Gadget was in the best of health and spirits, we have been able to put in two days of hard work and ceaseless activity. We visited the American consulate, and the Syndicat d'Initiative de Provence, the P.L.M. Railroad Information Bureau, Mr. Thomas Cook's office and various other joints, where we picked up all the information and statistics we could about Montpellier and its wine business, and about the other regions we expect to visit.

We procured guidebooks, road maps, books of traffic regulations, and a tremendous amount of miscellaneous information. Then we got number plates to put on the tractors, and licenses to drive them on the roads of France. This last was a terrible job; poor old Gadget had to talk and plead literally for hours with a half dozen official wise guys before she could convince them that we were all right and that our machines would not wreck the roads. After that we went over to the pier and claimed the three tractors and other machinery which had been unloaded from the boat. After getting this stuff through the customs, I put the big harvester in a warehouse, where I can send for it later. Everything else we are going to take with us. I hired enough help to get it all unpacked, set up, greased and stored in a garage down by the waterfront.

Then we hired a young French mechanic to drive the twenty-horsepower tractor. I am going to drive the sixty, and Gadget will drive the thirty.

The French mechanic's name is Jean. I am paying him seven hundred francs a week, which I think is a little more than thirty dollars, as near as I can figure this silly French money. When the proprietor of the garage heard about this, he pretty near fainted completely away. He took me and Gadget into his private office and explained that it would never do at all. He said that it was far too much. Jean would think we were soft, easy marks. He would lose all respect for us and would get exaggerated ideas about his own importance. He would become proud, haughty, impertinent, lazy and absolutely no good at all. At one and the same time we would be wasting our money and completely spoiling a good mechanic, who was a good worker and knew his proper station in life.

I had Gadget tell him that as I was an American, hiring a man to drive an American machine, it seemed reasonable to me to pay him American wages. "And," I said, "if he gets proud, haughty, impertinent or lazy with me, I will fire him right off that tractor onto his ear."

Then the garage proprietor said that we would be ruining not only Jean but all the other mechanics as well. "When they hear how much you are paying, they will all demand that I pay them just as much. It will cause me all kinds of trouble."

So I told Gadget to tell him that that was his hard luck, and that if he didn't like it he could just stuff his long whiskers inside of his mouth and chew on them for a while. And after Gadget had translated for him at least a part of what I had said, we walked out and found a sign painter's establishment, where we arranged for a lot of banners to give our tractor parade an artistic and dressy appearance.

Then we scouted around, and in a sort of junk store I was fortunate enough to find and purchase an apparatus which I am sure will not only cause the keenest pleasure among the inhabitants of the regions through which our tractor parade passes but will also make it absolutely certain that we shall attract a tremendous amount of attention. This apparatus is a second-hand steam piano, or calliope, complete with boiler of ample size. It was formerly the property of a small circus which went broke in this town some time ago, and I was able to pick it up for the ridiculous price of only a thousand francs. I have had this instrument placed in one of the dump wagons which we brought over from America, and I have hired a young musician—at about twice the usual emolument—to play upon it.

As well may be imagined, these activities kept us busy all day yesterday and today. We would have enjoyed another day to do a little sightseeing in the beautiful city of Marseilles, but Gadget and I are both so eager to get started on our great enterprise that we cannot consider any delay for this purpose. Consequently, bright and early tomorrow morning we are going to start forth on our sensational pilgrimage for the mutual benefit of the French nation and the Earthworm Tractor Company.

I will send you frequent letters regarding our progress. As you know, my address is in care of Thomas Cook, Marseilles.

Cordially yours,
ALEXANDER BOTTS.

ALEXANDER BOTTS
EUROPEAN REPRESENTATIVE
FOR THE
EARTHWORM TRACTOR

TARASCON, FRANCE.
FRIDAY NIGHT, MARCH 9, 1928.

MR. GILBERT HENDERSON,
EARTHWORM TRACTOR COMPANY,
EARTHWORM CITY, ILLINOIS.

DEAR HENDERSON: We have just completed the second day of our grand pilgrimage, and already we have learned of a possible sale. When we arrived in this interesting little town of Tarascon about two hours ago, we discovered a very interesting state of affairs which looks to me like an absolutely certain sale of at least one tractor. We have been spending the late afternoon and evening scouting and investigating—hovering watchfully above our prey. And tomorrow morning we hope to swoop down and with a short and snappy sales attack put over the first tractor sale in France. But before I relate the details of this interesting matter, I will give you a brief account of our trip.

Never in my life have I felt so proud and happy as I did yesterday morning, when our magnificent tractor parade debouched from the garage down by the Marseilles waterfront and swung magnificently up the Canebière. At the head of our column rolled the big sixty-horsepower machine, driven by myself; it was draped with the Stars and Stripes intertwined with the French Tricolor, and surmounted by a tremendous red-and-blue sign bearing the words "Ver de Terre." (Note: This is French and it means "Earthworm.") Hitched onto this first tractor I had our large twelve-foot blade grader and a wheel scraper. Behind this came the thirty-horsepower tractor, driven by Gadget herself, and bearing a large green-and-orange sign with the words "*Grand Tracteur Americain.*" (Note: This is French, and means "big American tractor.") This machine was pulling our large dump wagon, containing the plow and the fresnoes. Behind this was the little twenty-horsepower machine, driven by Jean, and bearing the inscription in small letters, "*Honneur et Gloire à la France et à l'Amerique.*" (Note: This is French, and it means "Hurray for France and America.")

This last machine was pulling the small dump wagon, which was of course decked with flags and which contained the tractor plow, the steam piano and the young French musician. Before we started, he had stoked up his boiler, opened the drafts and tied down the safety valve, so we had a truly terrific head of steam. When he started to bear down on the keyboard we got a burst of the loudest and most remarkable music I have ever heard. As we swept along that beautiful main street of the magnificent city of Marseilles, the entire region reverberated to the stirring and ear-splitting strains of the "Marseillaise," rendered as only a young and talented Frenchman can render it. I was so occupied with the driving that I did not have time to look around very much, but I feel sure we attracted a good deal of attention. In fact, I had a distinct feeling that the entire town was looking at us with the greatest awe and admiration.

On two or three occasions we were stopped by *les gendarmes.* (Note: This is a French word meaning "the cops.") Apparently, they objected to something, but as they talked only French and as I did not want to bother to ask Gadget to come forward as interpreter, I never did find out exactly what was eating them. Possibly they didn't like the noise. In each case, however, after I had talked to them awhile in English, they spread out their bands and raised their shoulders in a futile sort of a way and let me go on.

Note: This is the best way to handle these foreigners. Just talk to them in English, and in time they will give up.

It took us more than an hour to get out of Marseilles. We then followed the Route Nationale—that is French for "National Route"—through the hills, and by noon we had reached the flat country of the Rhône delta. The road was not bad at all. Of course it could not be compared with the new concrete highway between Chicago and Earthworm City, Illinois, but it was pretty good.

At one o'clock we stopped for a picnic lunch, and at once my French calliope player began to get a little temperamental. He had been alternately stoking the boiler and massaging the keys all morning. He remarked that he was tired of playing, and right away Jean said that he himself was twice as tired of listening. The young musician at once got very sore, and there was such an argument that I had to step in myself and quiet it. On the whole, I was rather inclined to sympathize with Jean. I find that the music of the steam piano—though it is wonderful at first—tends to become a trifle wearying after the first three or four hours. I tactfully suggested that the young musician might take things a bit easier through the afternoon, but he was so sore that he played practically continuously in the hope that it would annoy Jean. It did. And it also began to annoy me.

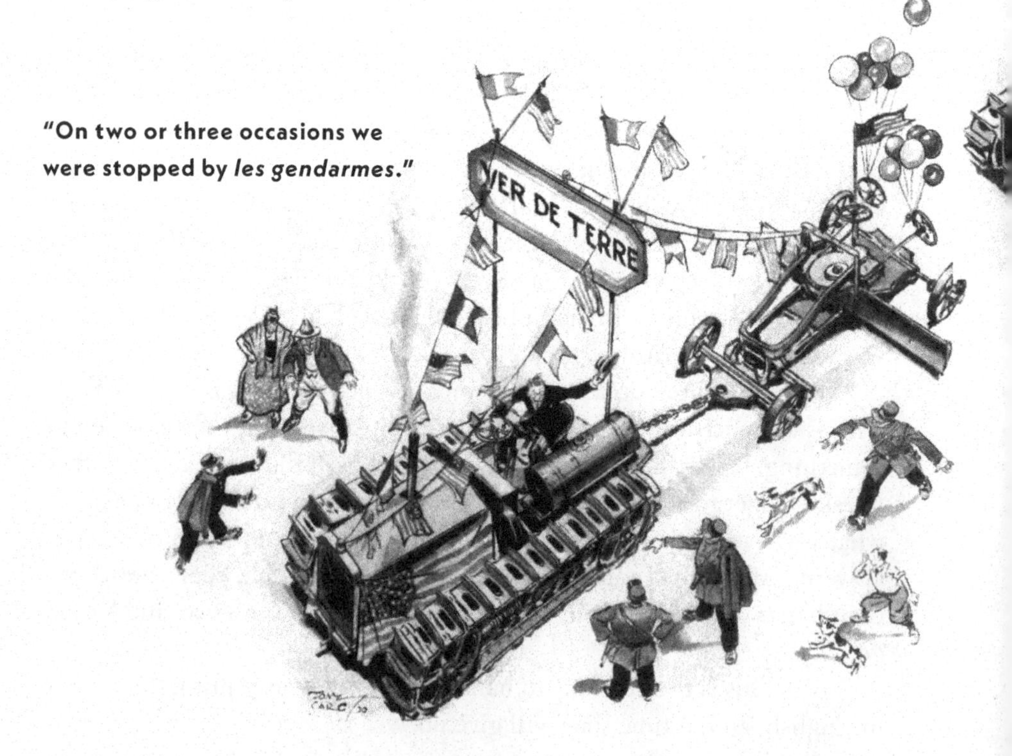

"On two or three occasions we were stopped by *les gendarmes*."

About suppertime we reached the small town of St.-Chamas, where I arranged for us to spend the night in a small inn. During the evening we showed the tractors and the machinery to a crowd of the inhabitants that had gathered around. Unfortunately, there did not seem to be any very lively prospects.

The next morning we got under way early. Before starting, I had Gadget tell the piano player very emphatically that he was to operate his instrument only at such times as I ordered, and he was not to keep up any such continuous uproar as he had the day before. This very reasonable request he seemed to regard as a reflection on the quality of his playing, and during most of the day's ride he sulked and grieved like a spoiled child. But he obeyed orders; we had music only while passing through towns, of which there were not many, and, in consequence, the rest of us had a much pleasanter time.

We had lunch at a place called Arles. And late in the afternoon, as we were approaching the town of Tarascon, we discovered something that filled my heart with joy. On the bank of the River Rhône, just south of Tarascon, we saw a large and splendid castle. It was built of stone. It had towers, battlements, a yard unrounded by tremendous double walls, a big gate, and a moat—just like the pictures in books. All around the castle were a lot of dingy, mean-looking houses and shacks.

All this was very interesting, but the thing that made me stop and prick up my ears was the fact that there seemed to be a lot of work going on. There were several dozen men hard at work tearing down some of the houses right next to the castle wall; others appeared to be grading the surface of the ground and smoothing it off at a place where some buildings had already been removed.

In other words, I saw that there was dirt moving going on. And wherever dirt is being moved, that is the place for me to step in and sell as many Earthworm tractors as the job can possibly use.

I did not stop at once. It was getting along toward the end of the afternoon, so I drove right on into the town, stored the tractors and the machinery in a garage, and engaged rooms in a hotel.

I then started out with Gadget to make inquiries. The first person we tackled was the concierge of the hotel. Fortunately, he spoke English.

Note: *Concierge* is a French word. The dictionary says it means "janitor," but most of these concierges seem to be a whole lot more than that. This bozo was very large and handsome, with big whiskers and a swell blue uniform with brass buttons. It seems that he speaks about four or five different languages. He greets the arriving and speeds the departing guests. He takes care of the baggage, is a general bureau of information, bosses the help, and handles most of the other jobs around the place. As far as the hotel is concerned, he seems to rate about like the captain on a boat. The only difference is that the concierge is more important and wears a much more handsome uniform.

This concierge was very glad to tell us all he knew about the castle. It seems that the name of it is the Château de Something-or-Other. It is very old; part of it was built by the Romans, and the rest of it in the thirteenth century—or maybe it was the eighth, or the sixteenth, or something. Anyway, it was built many years ago, and for a long time it was used as the home of some of the big nobility. At the time of the French Revolution it was taken over by the government, and for the last hundred years or so it has been used as a jail. That seemed to be about the only thing they could do with it. It was so damp and so disagreeable that nobody with any sense wanted to live there. Also, it was too big for any one family, and it would have been too much of a job to cut it into apartments. It was not located in the right place for stores, and of course the dampness made it useless for almost anything you could think of. In fact, the only people that could be persuaded to live there were the poor, helpless prisoners, and probably it took a good deal of heavy persuading to get most of them in.

"But recently," said the concierge, "the people of the town have got an entirely new idea—one that they think will bring a lot of money into the town. They are going to make this castle into a *monument historique*."

"And what," I asked, "is a *monument historique*?"

"It's a historic monument," said the concierge, "and it's a fine idea. They have tried it out at several towns near here, and it pays big money."

"How is that?"

"Well," said the concierge, "there is a town about twenty kilometers up the river called Avignon. There is a great big stone building up there—a regular whale of a building—that was used as a barracks for the soldiers. Apparently in the old days, whenever they had a big building that was too damp and unhealthy for anybody else to live in, they would use it for prisoners or soldiers, or, once in a while, as an orphan asylum. Soldiers and orphans and prisoners are all equally helpless. They have to live where their bosses tell them to."

"Yes," I said, "I was in the Army once myself."

"So the soldiers lived in this big building up at Avignon for a long, long time. It did not do them any good, and it didn't do anybody else any good. But some years ago one of the inhabitants of the town got to reading in a history book, and he found out that this building had once been the Palace of the Popes."

"I thought the popes lived in Rome," I said.

"They do now," said the concierge, "but it seems that a long time ago several of them lived at this little town of Avignon."

"And everybody in town forgot all about it until this guy happened to look in the history book?" I asked.

"Probably most of them knew that it had been the Palace of the Popes, but they didn't do anything about it until this man came along. He and some more people talked it over, and they decided it ought to be made into a *monument historique*."

"And just what," I asked, "do they do to an old building to make it into a *monument historique*?"

"They fix it up. They took this building at Avignon, and they got the soldiers out, and they repaired it and tried to make it look as much as possible the way it was in the old days. There had been a lot of painting on the walls which had been all whitewashed by the soldiers, and they got a smart guy down from Paris who managed to take off the whitewash without hurting the paintings underneath. I have been up there and looked at these paintings, and, personally, I would prefer the whitewash. But that is just because I do not know anything about art. Any picture that is very old and is painted in a clumsy way is art. The clumsier it is, the more artistic it is."

"Yes," I said, "I have seen something of this art myself, and I guess you are right."

"Of course I am," said the concierge. "And when they got the palace all fixed up, they hired a lot of guides. They charge admission, and every day they take a whole flock of tourists through the building. They make a lot of money. And that is not all."

"No?"

"There are so many tourists that they bring prosperity to the whole town. This is particularly true of the Americans. They stay at the hotels; they visit the shops; they spend money all over Avignon. It is a wonderful thing."

"It must be," I said.

"It is," said the concierge, "and the same pleasing condition exists at the town of Arles, a short distance down the river. At Arles they have dug up a whole lot of Roman ruins which, in former days, were nothing but a nuisance—cluttering up half of the town and occupying much valuable space that could have been used for useful buildings. But now, thanks to the Americans and other tourists, these ruins are the chief source of prosperity to the town. Naturally, this has all been very distasteful to the citizens of Tarascon. The tourists stop off at Avignon, throwing money right and left. Then they get into the train, motorbus or auto, and rush right through Tarascon to Arles, where they throw around a lot more money. They never stop here. It is most annoying."

"It must be."

"It is little short of a calamity. The citizens of Tarascon can use money just as well as the citizens of Avignon or Arles, and it is most maddening to see all this tourist wealth rolling through the town without stopping. But, fortunately, the situation is about to be remedied, for our old jail is now being converted into a *monument historique*, and when they get through with it, it ought to be one of the best *monuments historiques* in the whole country, and the beautiful city of Tarascon should then begin to get its rightful share of the tourist gold."

"This is indeed most interesting," I said. "Just how much work are they going to do in fixing up this old castle?"

"I do not know," said the concierge.

"Who has charge of the job?" I asked.

"It is under the direction of the keeper of the archives of the city."

"Then he is the man I want to see. What is his name and where does he live?"

"His name is Monsieur Jules Pascal," said the concierge, "and he lives on the rue du Rhône."

"Very well," I said, "I will call on him this evening. I thank you very much for your information."

"It is always a pleasure," said the concierge, "to be of service to the Americans; they are so very liberal."

So I slipped him a five-franc note, and Gadget and I went in and had a very good supper. After supper we called on Monsieur Jules Pascal on the rue du Rhône. He turned out to be a very pleasant and hospitable gentleman, and I guess he is a rather bright guy in spite of the fact that he speaks no English. Gadget got her French into action and talked with him for more than half an hour. Then we said good night and came back to the hotel, where Gadget has just been telling me what she learned.

It seems that there is a lot of work inside the building, which of course is of no interest to us. But there is also a lot of outdoor work. They are going to remove all the shacks and ugly little buildings that have been built around the great wall. They are going to clean out the moat, which has been largely filled in, and they are going to smooth up and regrade about fifteen or twenty acres of ground all around the castle. This will mean a whole lot of dirt moving. The job has been let out on contract to a man called Monsieur Alfred Beaumont, who apparently has no equipment other than picks, shovels and dump carts. The keeper of the archives did not know where Monsieur Beaumont lived, but he told Gadget that we could find him tomorrow at the château, where the work was going on.

Having learned all these facts, Gadget and I at once planned our sales campaign. Tomorrow morning we will descend upon this Beaumont guy with everything we have. It will be a quick, short and overwhelming attack. We will advance with our three tractors and all the equipment, including the steam piano. Without stopping to ask anyone's permission, we will drive right up to where the work is going on, order the workmen to one side, and get busy ourselves. With one tractor and a chain we will start pulling down the walls of the larger buildings. With a second tractor, which I have equipped with a heavy wooden pole to use as a battering ram, we will commence demolishing some of the smaller walls. And with the third tractor and the wheel scraper we will start leveling up the rough ground.

There will be no lengthy approach, no long arguments, no tedious explanation. Entirely without warning, the astonished Monsieur Beaumont will be treated to a tractor demonstration more amazing and more remarkable than anything he could possibly have dreamed or imagined. If he has any sense at all, he will realize at once that the Earthworm tractor is heaven's gift to dirt-moving contractors. And shortly thereafter I expect to have his name on the line and his cash in my possession.

I am so confident of the outcome that I have made a large bet with an American gentleman by the name of Jones, who is a buyer for an American silk company, and who is staying here at the hotel. I happened to mention to him what I was going to do, and he sprang the old pessimistic remark that you can't sell such things as tractors to the French. Right away I offered to bet him a thousand francs that before tomorrow night I would sell at least one tractor to a Frenchman. And he was foolish enough to take me up.

So tomorrow I will not only have the pleasure of making my first tractor sale in France but I will also rake in one thousand francs—which I am too lazy to figure out what it would be in real money, but which ought to amount to quite a bit.

I will write you again tomorrow and give you my final report on this interesting affair.

Most sincerely,
ALEXANDER BOTTS.

ALEXANDER BOTTS
EUROPEAN REPRESENTATIVE
FOR THE
EARTHWORM TRACTOR

TARASCON, FRANCE.
SATURDAY, MARCH 10, 1928.

MR. GILBERT HENDERSON,
EARTHWORM TRACTOR COMPANY,
EARTHWORM CITY, ILLINOIS.

DEAR HENDERSON: Things have been moving so fast today that for once in my life I find myself completely confused. I don't know whether I am coming in or going out. I don't know whether I have just put over one of the snappiest sales of my entire career or whether I have been played for a sucker. The only thing I am sure of is that I have lost the thousand-franc bet. I only lost because of a miserable technicality, but I lost; so I have paid Mr. Jones promptly and like a good sport.

Everything started off exactly as I had planned it. I carried through my high-powered selling campaign so efficiently and so rapidly that I never realized anything was wrong until after I had brought matters to what I thought was a successful conclusion.

Gadget, Jean, the musician and I had an early breakfast, and soon after eight o'clock we had the tractors and all the machinery out of the garage and ready to go. About this time Mr. Jones came out of the hotel. I invited him to come along, and he climbed up in the seat beside me. We started off in the same formation we had used on the grand march out of Marseilles. The sixty-horsepower Earthworm tractor, driven by myself, was at the head of the parade with the twelve-foot blade grader and the wheel scraper. Next came the thirty-horsepower machine, driven by Gadget, and pulling the large dump wagon containing the plow and the fresnoes. At the rear was the twenty-horsepower machine equipped with a big timber for use as a battering ram. This tractor was driven by Jean, and it pulled the small dump wagon containing the steam piano and the temperamental young French musician.

We swung down through the main street of Tarascon with the steam piano going full blast, much to the amazement and delight of the inhabitants. Within less than half an hour we had reached the castle, where I at once started to put on the demonstration with all my usual energy and skill.

In less than thirty seconds I had unhooked the large tractor, backed up to the wall of a half-ruined stone house, grabbed a large chain off the blade grader, passed it in one window and out another, hooked both ends to the drawbar of the tractor, and climbed back into my seat, ready to start up and pull down the whole wall. In the meantime, Gadget had driven over onto a piece of ground which had just been cleared of houses; it was all full of holes and hummocks, and Gadget started in right away with the wheel scraper to level the place off. At the same time, Jean—following previous instructions—had unhooked from the small dump wagon and advanced with his battering ram toward a rather thin little house wall that stood off at one side.

We had acted with such rapidity that the amazed workmen made no attempt to stop us. Most of them retreated as fast as they could to a safe distance and stood watching us with their mouths hanging open. I was just getting ready to slip in the clutch and pull down the big stone wall, when there came a sudden yell from the young French musician, followed by some frantic and discordant toots of the steam piano. Glancing up, I observed a very curious phenomenon. When Jean had unhooked from the small dump wagon he had left it on top of a little knoll, and now it had started to roll down the slope toward the castle moat, coasting along quite merrily, and carrying with it, of course, the steam piano and the piano player. This piano player, besides being temperamental, also seems to be a good deal of a dumbbell. Instead of putting on the brake, be merely sat in his seat, yelling and pounding on the keys.

With my usual presence of mind, I leaped off the tractor, rushed over, and pulled the brake lever on the side of the wagon. This stopped the wagon, but not in exactly the way I had intended. In the hurry and excitement of the moment I quite naturally grabbed the first lever I could reach. And most unfortunately—but, of course, through no fault of my own—I got hold of the dump lever instead or the brake lever. The doors in the bottom of the wagon at once opened, and the steam piano and all its accessories, including the boiler and the piano player, dropped through and hit the ground with a very healthy smack. This blocked the wheels and prevented the wagon from rolling into the moat. But it did not seem to be very good for the steam piano. Several of the pipes broke off and a great cloud of steam came hissing forth. And the mishap seemed greatly to excite the piano player. He came scrambling out of the wreckage and began to act very temperamental indeed, waving his arms about and talking French so fast that I did not think even a Frenchman could understand him.

But Gadget—as I have told you before—really knows more French than the average Frenchman. She came running over and was able to interpret the gist of his remarks. He said that never in his life had he been treated so outrageously. He was an artist. He had been putting his whole soul into the music which he was producing. And what reward did he get? Nothing at all. The first two days he had been forced to listen to the vulgar and ignorant criticisms of Jean. For two days he had been constantly insulted. And this morning he had been brutally assaulted. He would stand no more. He would quit at once and go back to Marseilles.

This last remark seemed to me the most sensible one we had heard from this musician so far. The steam piano seemed to be out of commission. If it were beyond repair, we would of course have no use for our musical artist. On the other hand, if it proved possible to repair the instrument, it would be even more desirable to fire this guy so that we could save our eardrums and our nerves from his overenthusiastic and needlessly continuous music. I therefore had Gadget hand him a week's salary, with enough extra to buy a ticket back to Marseilles.

As we were wishing the young man a thankful farewell, we were interrupted by a large and tough-looking gentleman who came striding up and addressed us in English.

"Well, well" he said. "You seem to be Americans. What do you think you're doing? Where did you get all this machinery? Who gave you permission to come in here anyway?"

"Who are you?" I asked.

"My name is Beaumont," he said. "I am in charge of the work here. And it seems to me you have a good deal of nerve. What do you mean by butting in on my job this way? What's the big idea?"

As soon as I realized that this was the big boss himself, I started to work on him with all my usual energy. Unfortunately, I was still a little confused on account of the accident to the steam piano, and for this reason I fear that I did not handle the affair with all the wisdom and good judgment that I usually show.

I at once shook hands most cordially. "It is indeed a pleasure to meet you, Monsieur Beaumont," I said. "My name is Alexander Botts. I am the European representative of the great Earthworm tractor—that world-famous miracle of science and machines. The three tractors and all the accessory machinery which you see before you are for sale. I am about to put on a demonstration which will show you that this is exactly the stuff that you need for your work here."

"Let me get this straight. You say you want to sell all these tractors and all this machinery?"

"It is for sale in whole or in part. I will now start the demonstration."

"I don't want to see any demonstration."

"But you must. Let me show you what I can do. I am going to pull down that stone wall over there behind the big machine."

"You certainly are not," said Monsieur Beaumont. "That wall is one of the ancient outworks of the castle. If I let you pull it down, the authorities would pretty near murder both of us."

At this point, I realized that I had almost made a faux pas. (Note: This is French and means a "bad break.") It occurred to me that perhaps I was not making the best possible impression on Monsieur Beaumont. I began casting about for some soft answer to turn away his possible wrath, when he surprised me by asking the price of my entire outfit. Now, I rather dislike to discuss prices at an early stage of a selling campaign, before I have had an opportunity to kindle a warm desire for the goods in the mind of the prospect. But Monsieur Beaumont asked the question with such brutal directness that I had no choice but to answer. On the back of an envelope I rapidly added up the prices of the three tractors, the grader, the wheel scraper, the two wagons, the plow, and the two fresnoes

"The lot for 459,250 francs," I said. "And I will throw in what is left of the steam piano, free of charge."

"Sold," said Monsieur Beaumont.

"Pardon me?" I said.

"Sold!" he repeated.

For a short time I was so amazed by the turn that events had taken that I did not know whether I was on foot or in an airplane—whether I was actually engaged in a selling deal or merely reading a fairy story. I had supposed that this man would be a difficult prospect. I had expected that it would be a difficult job to sell him even one tractor. And here he wanted to buy the whole works. Furthermore, he had closed the deal quicker than any prospect I had ever worked on.

"But I don't understand," I said, and I fear that my face must have presented a very stupid appearance. "Don't you want me to tell you something about these machines? Don't you want to see them work?"

"No," said Monsieur Beaumont, "I can see that they are just what I want, so we might as well get down to business at once. Come with me."

He led me over to his automobile, which was parked nearby, and drove me down to the bank in Tarascon. Here I made him out a bill of sale for the entire outfit at the price I had mentioned, and he handed me the full amount in beautifully colored French notes. I was still in such a daze that all I could do was thank him in a confused sort of way and stuff the money in my pocket. Then we drove back to the castle. By the time we arrived, I had partially regained consciousness, and I began to think that I had done something pretty good.

Monsieur Beaumont wanted to hire my mechanic Jean, so I introduced them, and then rushed over to Gadget and Mr. Jones.

"Mr. Jones," I said, "you owe me a thousand francs. I have not only sold a tractor, I have sold three tractors, and all the machinery as well. Congratulate me and pay me."

"Not a chance," said Mr. Jones. "You owe me a thousand francs."

"How so?"

"You bet me you would sell a tractor to a Frenchman. And that man is not a Frenchman. He is an American. While you were gone we talked to the workmen here, and they say that Mr. Beaumont is an American contractor."

"I don't believe it," I said.

"All right, I'll prove it to you," said Mr. Jones. He called Monsieur Beaumont over, and Monsieur Beaumont told us that he was indeed an American. He had been a big contractor in Kansas City. While taking a pleasure trip through France, his wife had been taken ill and was now in a hospital at Nice. He did not want to go back to America until his wife was able to accompany him. The life at Nice bored him, so he had left his wife and daughter there, and was amusing himself by taking a little flyer in the dirt-moving business in France. His French-sounding name came from a French Canadian great-grandfather, and he was just as much of an American as anybody else.

So I had lost my bet. And that is not all. Right away Gadget began to jump on me.

"Your job over here," she said, "is to introduce these tractors to the French people. This sale to an American doesn't really count at all. And now that you have got rid of all your demonstration machinery, we are sunk, as far as France is concerned, until we can get more shipped over. We ought to be going on to Montpellier and getting after the real Frenchmen. But now we can't."

By this time, I had fully regained consciousness after the jolt I had received by making such a quick sale. And I began to realize that I had conducted this selling deal not wisely but too well. My first move was to chase after Mr. Beaumont and do something I have never done before in all my years as a tractor salesman. After selling a number of tractors, I turned right around and made a frantic effort to buy back at least one of them. But I had no luck at all. Probably I am so used to talking stubborn bozos into buying tractors that I cannot accommodate myself to the reverse procedure of talking them into selling tractors. Mr. Beaumont only laughed at me. He said he had used Earthworm tractors around Kansas City and he knew what wonderful machines they were. That was why he had bought them so quick. And that was why he was going to hang on to them.

Finally I had to give up, and Mr. Beaumont drove us back to the hotel, where it has just occurred to me that the whole trouble is that I am too good a salesman. I just can't seem to resist making sales, even when I shouldn't.

I have just cabled the New York office to ship a new fleet of demonstration machines to Marseilles. While awaiting their arrival, Gadget and I will make a whirlwind raid into Italian territory. I have just received a letter, forwarded via Marseilles, from a man in Merano, Italy, who has seen some of the advertising literature which we sent out. He wishes to see a demonstration, and he says that if the machine lives up to the claims of the advertisement, he will buy. Accordingly, we are leaving for Italy tonight. You will remember that I still have four demonstration machines at Genoa. I will have one of these sent up to Merano, where I hope to put over a real bona fide sale to a real European.

And it gives me the greatest pleasure to announce that I will have a certain amount of extra expense money for this Italian trip. In checking things up, I find that I was so excited and confused when I made the sale to this Beaumont man, that I made a mistake in addition, and charged him twelve thousand francs too much. As he seems to be so satisfied, and has been so snooty about taking back anything connected with that sales contract, I have decided not to refund any of this money. I have used a thousand francs to pay my bet to Mr. Jones. Another thousand will reimburse us for the steam piano. And the remaining ten thousand will be used to make our Italian sales campaign even more magnificent than our French venture. I sincerely hope that before long I shall be able to send you tidings of a glorious victory in sunny Italy.

Sincerely,
ALEXANDER BOTTS.

THE VINEYARD AT SCHLOSS RAMSBURG

ILLUSTRATED BY TONY SARG
OPENING ILLUSTRATION BY NICK HARRIS

ALEXANDER BOTTS
EUROPEAN REPRESENTATIVE
FOR THE
EARTHWORM TRACTOR

HOTEL FRAU EMMA,
MERANO, ITALY.
WEDNESDAY EVENING,
MARCH 21, 1928.

MR. GILBERT HENDERSON,
EARTHWORM TRACTOR COMPANY,
EARTHWORM CITY, ILLINOIS.

DEAR HENDERSON: It is with a feeling of the greatest optimism that I write you to report that I have just arrived in this charming city along with Mrs. Botts—or Gadget, as I usually call her—and that we are about to start the great selling campaign which we feel sure will proceed from one success to another until at last there will be a real American Earthworm tractor on every farm, every dirt-moving project, and every hauling job in the entire kingdom of Italy, from the southernmost shore of sunny Sicily to the northernmost pass in the snow-covered Alps.

Right now we are in the Alps. I would like to write a whole book about this place, but as I do not wish to be prolix I will merely state that the town of Merano is in a deep valley at the confluence of two rushing mountain torrents. There are splendid hotels and stores and hundreds of ancient stone villas with beautiful gardens and parks and avenues of trees—very fine to look at in the first green of the springtime. The sides of the valley, which rise up very steep all around the town, are covered with vineyards. There are grim-looking medieval castles on a number of the lower hills. And behind these are the dizzy peaks and snowfields of the Tyrolean Alps. In short, this is indeed a swell place, and in every way worthy to be the scene of our first Earthworm-tractor sale in Italy.

Our ten-horsepower demonstration machine has already arrived by freight from Genoa, and tomorrow morning we are going to call on Graf Anton Hasendorf, who lives in a nearby castle called Schloss Ramsburg. (Note: *Graf* is German for "count," and *Schloss* means "castle".) This Count sent me a letter in German—which Gadget, who knows all these languages, translated—saying that he had seen one of our advertisements and would buy a ten-horsepower Earthworm provided it was as good as the advertisement claimed. This guy uses the German language because he is an Austrian count. Apparently this region used to belong to Austria. It didn't become part of Italy until after the war, and most of the people here speak German. All of which makes no difference to us, as Gadget is such a splendid linguist that she can talk to any of them.

I will write you again tomorrow night and let you know how we get along. But don't worry; I have a feeling that this Count is going to be as putty in our hands.

Most sincerely,
ALEXANDER BOTTS.

ALEXANDER BOTTS
EUROPEAN REPRESENTATIVE
FOR THE
EARTHWORM TRACTOR

HOTEL FRAU EMMA,
MERANO, ITALY.
THURSDAY EVENING,
MARCH 22, 1928.

MR. GILBERT HENDERSON,
EARTHWORM TRACTOR COMPANY,
EARTHWORM CITY, ILLINOIS.

DEAR HENDERSON: We have called on Count Hasendorf, and it has been a very hard day. I will have to admit that I am amazed and bewildered, and almost ready to give up trying to sell this gentleman a tractor. I will give you a brief account of what has occurred, and in the course of the narrative I will make a few remarks about the advertising department of the Earthworm Tractor Company, and the truly remarkable way they seem to translate their publicity into foreign language.

The beginning of our day's work was entirely favorable. We had already inquired the way to the castle; so, after an early breakfast, we got the tractor at the freight station, filled it with gas and oil, and climbed on board.

With the gears in high, and with the clear morning sunshine playing on the brand-new paint of the tractor, we rolled majestically down the Corso Goethe, waking all of the guests in the swell hotels along the way with the cheery, whole-souled roar of our motor. At the municipal theater we turned to the right, crossed the bridge over the swift, glacier-fed waters of the Passirio, and followed the Viale di Campi past the Camp Sportivo—where our advent temporarily interrupted several early-morning games of tennis.

A mile or so farther on we passed the bottom of a beautiful new funicular railway, if that is what you call it—two big cables stretched right through the air from the bottom of the valley to a town called Avelengo about three thousand feet up on the mountainside. A little car, hung on rollers, and pulled by a smaller cable, was running up one of these big cables, and a second car was coming down the other. There were passengers in both cars. Right beside the splendid new funicular was an older one, which had apparently been abandoned.

It seemed to me that all this was a good omen. If the people of this region used such magnificent transportation equipment, they ought to be machinery-minded and apt to look with favor on that greatest of mechanical masterpieces, the Earthworm tractor.

As we continued on our way the road began to climb, and directly ahead and far above us we saw Schloss Ramsburg itself, perched on top of a rocky hill or promontory. The road zigzagged up the side of the valley, circled around behind the castle, and then took us out along a narrow ridge that connected the main range of mountains with the hill on which the castle was built. At length we came to a deep gully which was spanned by a drawbridge leading to a big gate in an ancient stone wall. We had arrived.

We parked the tractor beside the road, walked across the bridge, and rang a large bell. I expected to see a guy in armor come out, but such was not the case. A little door in the big gate opened, and a very harmless-looking old woman greeted us. Apparently she was the cook or housemaid or something.

Gadget told her in German that we wanted to see the boss, and she led us inside. We walked through a couple of arches and came out into one of the best-looking courtyards I have ever seen. It was paved with big blocks of stone and surrounded by stone buildings with beautiful arched doors and windows, tile roofs, towers, gables and balconies. One of the buildings, at the lower end of the court, seemed to be a sort of combination stable and servants' quarters. In this country the servants and the animals rank about the same. In front of the stable was a batch of laundry drying on a clothesline, and a large and very rich manure pile with a flock of fat hens scratching around in it. All this looked most domestic and homelike. And the whole place, taking it by and large, was undoubtedly a swell joint. All the stonework was very old and covered with lichens and moss and ivy. Very fine, indeed.

The old woman took us through an archway at the opposite end of the court from the stable, and we entered the great hall of Schloss Ramsburg, which was filled with a warm glow from a cheerful fire in the big fireplace. I had just time to notice the magnificent Turkish rug on the stone floor, the rows of distinguished-looking old portraits on the wall, and the huge beams overhead, when Count Hasendorf himself came in.

He was better than any of the portraits—about six feet tall, about seventy-five years old and with a face and a beard like the pictures of old man Moses. Beside him was the Countess Hasendorf, a quiet, gentle-looking woman with very lovely white hair. After one look I knew that these people were the real stuff. And I congratulated myself for having been sensible enough to bring my wife with me. For Gadget is the real stuff too. Gadget at once addressed them—in German, of course—and they replied in the same language. We learned later that they speak German, French and Italian, but practically no English.

After the preliminary greetings were over, Gadget said, "We have brought an Earthworm tractor for you to look at."

"That is very kind of you," said the Count.

"We understand," said Gadget, "that you want to buy one."

"Well," said the Count, "I can't say that I really want to buy one."

"But I don't understand. I thought you wrote us that you would take a tractor if it was as good as the advertisement said it was."

"Exactly," said the Count. "I may be compelled to buy one of your machines. But I don't want to. I am an old-fashioned man, Mrs. Botts, and I hate this frightful modern machinery. Every fiber of my being is outraged at the thought of my beautiful vineyard being polluted by the presence of a vile tractor, clanking end puffing among the vines, destroying the peace of the whole countryside with its hellish noise, and polluting the very air with its foul exhaust. And when I consider the peasants who work for me—simple, sturdy souls whom I love as if they were my children, and who have always lived clean lives close to Nature—I am deeply distressed at the thought of their being tainted by contact with the soulless machine-made civilization of the cities.

"But these are evil days; I have encountered a difficulty in my vineyard operations: I have apparently lost a valuable mule; and it may be that the only way I shall be able to solve my problems will be by the purchase of a tractor."

"Well, Count," said Gadget, "that certainly is tough luck. But after you have bought an Earthworm tractor, you'll find out it's not half as bad as you think. Let's go out and look at the machine. Then you can tell us about your problems, and we'll see if the Earthworm can solve them." And she led the way back through the courtyard to where we had parked the tractor.

Note: The above conversation, as I have said, was all in German. But as you guys at the home office would naturally be too ignorant to understand it in the original, I have put it down in English just as Gadget translated it. And here is another idea: I had always thought of German as essentially a comic language—it has so many silly gurgles and snorts in it. But the way the Count talks it, it sounds most dignified. And the way Gadget handles it, it is actually musical.

When we got out to the tractor, the Count and the Countess sort of hung back.

"Step right up, Count," said Gadget. "It won't bite you. It's housebroken and everything. And its morals are clean and spotless—even if it does come from the materialistic machine-made civilization of the cities. It is guaranteed not to taint or sully the purity of even the most innocent Tyrolean peasant."

But the Count wouldn't come within ten feet of it. "Suppose you drive it in," he said, "and I'll show you what I want you to do with it."

As soon as Gadget had explained this to me, I flipped the crank, climbed into the seat, and drove across the bridge, through the big gate and into the courtyard. Then the Count directed me through another gate and out onto an open stone terrace on the side of the castle toward the valley, where we got one of the swellest views I have ever seen.

In front, and almost a thousand feet below us, was the town of Merano, with its churches, beautiful buildings, parks and trees. The two little streams—the Passirio and the Adige—glittered in the sunshine as they flowed along through the green fields and past a number of clustered little villages. On the hills across the valley were four or five big castles; the largest of which was Schloss Tirol, from which this whole region takes its name. And above the castles were the snow-covered peaks of the higher mountains.

After we had admired this beautiful scene, the Count explained all about his vineyard, and the troubles he was having, and what he wanted the tractor to do. As the conformation of the vineyard is the basis of most of the Count's troubles—and of all of mine at this time—I will describe it in some detail.

As I have said, Schloss Ramsburg stands on top of a hill which juts out from the side of the valley much as a buttress juts out from the side of a building. The only road to the castle—the one we have followed—winds up the slope of the main mountain range at one side of the hill, circles around, and reaches the castle from the rear, where the little gully, crossed by the bridge just outside the big gate, is the only thing that separates the top of the hill from the main slope of the mountain behind.

The right and left sides of the buttress-like hill are almost perpendicular rock cliffs, but in front of the castle the ground slopes down at a forty-five-degree angle for several hundred feet, and then continues by a gentle descent to the bottom of the valley and the town of Merano. On the upper, or steep, part of the slope is the castle vineyard. It is terraced like a gigantic flight of steps; and each step, or terrace, is a level shelf of rich soil about a quarter of a mile long and ten feet wide, held in place by a vertical stone retaining wall about ten feet high. The grapevines grow on trellises made of wooden poles.

As we looked down from the castle on this remarkable vineyard, the Count explained that it was one of the finest in the world.

"It was started by the ancient Romans," he said. "Possibly the stone walls that hold the terrace were begun two thousand years ago. It must have taken centuries to complete them—there are forty-four terraces in all—but apparently they were all finished when my family came into possession of the land in the year 1433. And the retaining walls are so solid that hardly any repair work is needed—which is fortunate, because with labor as expensive as it is today, and the price of wine so low, it would be impossible for me to do much stone construction."

"It looks like more work than it's worth," said Gadget. "Why couldn't the old Romans have put their vineyard on some level field?"

"A hillside vineyard is always better," said the Count. "The air drainage protects the vines from the danger of frost and from the dampness which causes fungous diseases. And the drainage of the soil water keeps the earth moist, but never water-soaked. Furthermore, the soil in this particular slope is of a chemical composition peculiarly adapted to grapes."

"How interesting!" said Gadget.

"History records," said the Count, "that the wine from this vineyard was the favorite drink of the great Emperor Augustus Caesar himself. From his time on, down through the ages, it has remained the first choice of the greatest authorities."

"I see," said Gadget. "It's a very high-grade product, like champagne."

"The wines of Champagne," said the Count, with the greatest contempt, "are mere trick beverages, filled with silly bubbles for the amusement of fools and the newly rich. The man of discernment wants flavor, not an amusement of soda water. The real connoisseur will use the products of Bordeaux, of Burgundy, perhaps even the famous Schloss Johannisberg on the Rhine. They are all good, but they do not completely satisfy him. And then, if he is fortunate, he discovers Schloss Ramsburg, and he knows that he has found the perfect wine, with a flavor, aroma and bouquet absolutely unique."

"You don't know how you impress me," said Gadget. "I didn't realize how famous the vineyard is."

"It is not exactly famous," said the Count. "It is, in fact, practically unknown to the vulgar multitude. The vineyard which you see before you contains the only soil in the whole world capable of producing our heavenly nectar. There is very little of it, and it is therefore unknown to all except a chosen few."

"I understand you exactly," said Gadget. "And now suppose we get down to business. I think you said you were having some troubles in operating this splendid vineyard. You lost a mule or something."

"Yes," said the Count. "I will explain the whole matter. You see, to get the best results, we must cultivate the vineyard very thoroughly."

"Naturally."

"And in former times this was easy. Labor was very cheap and the work was all done by hand—that is, with spades and hoes. But when I inherited the property about twenty years ago, the cost of labor had risen to some extent, and I decided to do the work with mules."

"Very sensible of you."

"But it was a difficult thing to manage. The vineyard was made for handwork only. There are no roads in it. The only way to get from one terrace to another is by means of steps. Come; I will show you."

The Count led us down a short slope, through a gate, and out onto the uppermost terrace of the vineyard. We walked under the spreading trellis to the other side of the ten-foot strip of soil and looked down over the stone retaining wall to the next terrace, ten feet below. Running along the face of the wall was a steep flight of rough stone steps not more than a foot and a half wide.

"That," said the Count, "is the type of steps we have all the way down the hill; perfectly adequate for the workman, but not much good for animals or wagons."

"I should think not," said Gadget.

"At first," continued the Count, "I despaired of ever being able to use mules here. But finally I discovered among my peasants an old man who seemed to be a genius in training animals. In the course of a single summer, by the use of infinite patience and endless hard work, he succeeded in training six young mules to walk up and down these steps as easily and as readily as a man."

"It seems incredible," said Gadget. "The steps are so narrow and so steep."

"I know they are," said the Count, "but this man did it. And later, when some of the first mules died, he trained others to take their places."

"So that solved your problem?"

"It did for a while. But five years ago the old man died. And last fall the last of the mules which he trained also passed away. And we can't train anymore."

"If it was done once," said Gadget, "you ought to be able to do it again."

"So I thought. But I have just about given up hope. The son of the old man has tried again and again. He thinks he is using the same methods, but something must be lacking. Perhaps he does not have the angelic patience, or the sympathetic understanding of mules that was possessed by his genius of a father. At any rate, he has failed. I have hired professional horsebreakers and animal trainers from all parts of Europe. And they have failed."

"Absolutely no luck at all?"

"Well, we had one partial success—which has only made matters worse. Last winter a horse trainer by the name of Antonio da Vinci came up here from Florence. He claimed to be descended from Leonardo da Vinci, but he had none of the great qualities of his alleged ancestor. I let him try his skill with one of the best mules I have ever owned—a splendid, vigorous animal by the name of Brunhilde. He started in at the bottom of the hill, and succeeded in leading the poor creature up over twelve of the walls—the steps down there are a little wider than those higher up—but when he came to the thirteenth wall, Brunhilde would go no farther. He

"Brunhilde would go no farther. Da Vinci used persuasion."

used persuasion. He used the whip until I had to order him to stop. He got ropes, and with fifteen men to push and pull, he tried to get that animal up those steps. But it was no use."

"So he finally had to lead her down, again?"

"Ah, no. He couldn't do that either. You see, it is harder for a mule to go down steps than up. And Brunhilde absolutely refused to go down. She would go neither up nor down."

"You mean she's still there?"

"For the past three months Brunhilde has been living on Terrace Number Twelve. You see that little patch of white down there?"

Far beneath us, near the bottom of the vineyard, was a little white square object.

"That," said the Count, "is a tent we put up to protect the unfortunate beast from the weather. And there she lives in idleness while we ought to be using her to haul wood down from the mountain. This summer we can use her to cultivate Terrace Number Twelve, and that is all. Unless we find a way out of this difficulty, we shall lose most of her value. It is pitiful."

"Isn't there any way to get on and off these terraces from the ends?"

"The terraces terminate at both ends in rocky cliffs."

"Couldn't you build ramps of some kind which mules could walk up and down?"

"It would be too expensive. Remember, there are forty-four terraces."

"It seems like quite a problem," said Gadget.

"I have been very much worried about it," agreed the Count. "I have worried about the mule, and also about the cultivation of the vineyard. Labor is now far more expensive than it was when I first began to use mules. If I had to go back to the old hand methods of cultivation, I would surely go bankrupt. Up to a few weeks ago I was almost in despair. And then I happened to see your advertisement down at the Merano Chamber of Commerce, and it occurred to me that one of your tractors would solve all my difficulties."

"Well, that's interesting," said Gadget. "But just what was your idea? Exactly what were you going to do with the tractor?"

"I thought we would drive it down into the vineyard, load the mule onto it and drive out again. After that we would keep the mules out altogether, and cultivate the whole thing with the tractor. You can cultivate a vineyard all right with a tractor, can't you?"

"Sure we can," said Gadget. "That is, we could cultivate any one of those terraces if we once got onto it. But what bothers me is this driving in and out that you speak of."

"That ought to be easy enough," said the Count.

"But those steps aren't wide enough."

"Then you could drive your machine right up and down the walls."

"Pardon me?"

"I say you could drive right up and down the faces of the walls."

"I'd certainly hate to try that."

"You needn't be afraid of hurting the walls. They are so solid that it wouldn't hurt them at all."

"No, probably not. But suppose you just wait a minute, Count. I've got to discuss this with my husband."

Gadget at once went into conference with me in English, explaining exactly what the old guy wanted.

"Let me get this straight," I said. "He wants to drive the tractor right down over the tops of those stone walls?"

"Not only that," said Gadget, "after he has driven down, he wants to drive up again."

"He's full of bananas," I said. "What does he think a tractor is—an airship, or some sort of a steel grasshopper or something? Are you sure that's what he wants to do?"

"That's what he said."

"Gadget," I said, "you are a wonderful linguist, and nobody admires you more than I do, but this time you have pulled a boner. You have misunderstood this guy."

She turned back to the Count. "Possibly I didn't quite understand you," she said. "Did you really think a tractor could be driven up and down those walls?"

"Certainly, it looked to me like a difficult feat, but I know very little about this modern machinery. And the advertisement said it could be done."

"What advertisement?"

"The one which your company sent out, and which the Merano Chamber of Commerce was kind enough to turn over to me. I have it here in my pocket."

"Would you just as soon let me see it?"

"With pleasure."

The Count reached inside his coat and handed Gadget a small folder. Gadget went into conference with me for the second time. We looked over the pamphlet.

And right at this point I want to interrupt my narrative and offer a little mild criticism of the advertising department of the Earthworm Tractor Company. Of all the low-grade morons that ever foisted their alleged services on a trustful and unfortunate corporation, those babies in the so-called advertising departments are the most amazing.

In the first place, this ad they sent the Count is in French. Can you tie that? Apparently the stenographers in the advertising department think that if a guy is a foreigner, all they have to do is send him an ad in some foreign language—any foreign language will do. So they send a French pamphlet to a part of Italy where everybody speaks German. But let that pass. We once sent some literature in English to New York City, and they finally found somebody that could translate it into Yiddish for them. And in this case the Count knows French, so he could understand what it said. In some ways it would have been better if he couldn't have understood it.

Because, in the second place, it was translated in a completely idiotic manner. And in the third place, the English pamphlet from which it was translated was no good in the first place.

I have a copy of the English edition with me. It is called, "The Earthworm in the Mountains," and it describes the use of tractors in hauling freight over the high passes of the Sierras and the Rockies. Among other things it says, "The Earthworm, because of its marvelous claw-like steel grousers, is easily able to climb the highest and most difficult of the Rockies," referring, I suppose, to the fact that an Earthworm has actually been to the tops of such mountains as Pike's Peak. Now this sentence, although somewhat idiotic, would probably be fairly harmless in the United States. But what does the advertising department do? It hires some boob of a professor of romantic languages, and he translates it into French as follows: "*Le Ver de Terre, à cause de ses ongles d'acier merveilleux, sait grimper facilement sur les rochers les plus hauts et les plus difficiles.*"

Probably this dizzy transistor thought he was doing all right to translate "Rockies" into the French word "*rochers.*" But "*rocher,*" although it is from the same root as our word "rock," is used to refer to a cliff or precipice made of rock. So the French advertisement really says, "The Earthworm, because of its marvelous steel claws—or toenails—is able to climb with ease up the highest and most difficult cliffs."

When Gadget had explained all this to me, I was, naturally, much disturbed. For the moment I was even what you might call nonplussed. For it was evident that Count Hasendorf had acted in entire good faith. He was, by his own admission, completely ignorant of modern machinery and had no way of judging what a tractor ought or ought not to do. When he read that the Earthworm could climb right up the face of a cliff or a precipice, he was surprised—just as he had no doubt been surprised at the performance of airplanes and the radio—but as long as the advertisement claimed this feat of climbing could be done, he assumed that we could back up our claim. If we could climb a cliff, he reasoned, we could negotiate a little ten-foot wall with no trouble at all.

"Oh, that advertising department!" said Gadget, as we considered the situation. "I would like to put my foot in their face."

"When we get back to the United States," I replied, "we will do exactly that—that, and much more. But for the present we must consider what we are to do here and now."

We considered. We decided on a course of action. And Gadget once more addressed the Count.

"Count," she said, "we want to tell you about the Earthworm Tractor Company. It's the finest company in the world. The departments which handle the engineering, design, production, service, and particularly sales, are made up of men of the highest intelligence. But unfortunately, as Shakespeare says, there may occasionally be something rotten even in so fine a kingdom as Denmark. And this is the case with our splendid company. In the interest of truth we must admit it, much as it shames us."

"I don't understand," said the Count.

"It's like this," said Gadget: "Through a mistaken sense of charity, the Earthworm management has seen fit to hire for our advertising department a small crew of pathetic incompetents, who, except for this aid, would no doubt be totally unable to gain a livelihood, and would, in consequence, be on the town or confined in some institution for the feebleminded. This group of alleged advertising men have produced, in their ignorance and folly, the unfortunate little booklet which you have shown me. And it is my painful duty to tell you that the statements in this booklet are largely applesauce and broccoli. There is no tractor built, not even the Earthworm, which can climb up or down vertical stone walls much as these in your vineyard, let alone wander around over cliffs like a fly on the wall. The advertisement was in error. We admit it, and we apologize."

Note: This speech was all in German, of course. I have given you an English rendering so you can see whet a splendid talker Gadget is, and so you can show it, with my compliments, to those poor fish down in the advertising department.

"This is indeed too bad," said the Count when Gadget had finished speaking. "I had hoped that your tractor could help me out of my difficulties. But I now see that it is impossible. I am sorry that you have had this long trip to Merano to no purpose. And I sincerely hope that we may part friends."

"Not so fast, Count," said Gadget. "Just keep your shirt on."

"Keep my shirt on?"

"Absolutely. We aren't licked yet by any means."

"But what can we do?"

"To tell you the truth, Count, I don't know. But we have come many miles to sell you a tractor, and we don't like to leave without doing it. So, if you don't mind, we'd like to look this proposition over for a few days and see if we can't work out some system for using our machine in this vineyard."

"I should be very glad to have you," said the Count. "My wife and I are leaving tonight for a few days in Vienna, but I will leave word with my men that you are to be given every opportunity to look over the vineyard and investigate our needs."

"That's fine, Count," said Gadget. "And can we leave the tractor here?"

"Certainly," said the Count.

We walked back into the castle, and I ran the machine into a stone shed at one side of the courtyard. Then Gadget and I said goodbye to the Count and the Countess, and walked down through the vineyard.

It was sure a tough-looking terrain for a tractor. I tried to think of some practical scheme for getting from one level to the next, and there just didn't seem to be any answer. It might be possible to break a hole in each one of those forty-four walls, and grade down the dirt so as to make a slope the machine could have climbed. But it would be too expensive. I thought of making a whole set of wooden ramps. Again, too expensive. I considered a single portable wooden ramp. But if I made it strong and solid enough, it would be too clumsy to move. A portable crane to lift the machine up and down seemed out of the question. It was most discouraging.

We walked down from one terrace to the next, using the little narrow stone steps, until we came to the terrace where the tent had been pitched. As we passed the tent, Brunhilde looked out at us with a very sorrowful expression. Evidently she didn't like being marooned in this way. We had a lot of sympathy for poor old Brunhilde—we felt pretty sorrowful ourselves. After petting her a bit, we wished her good luck and continued on our way to the bottom of the vineyard.

From here a short walk took us to a trolley line which brought us back to the hotel in time for lunch. This afternoon we spent wandering around the town, listening to the band on the promenade, and having afternoon tea at the Casino, or *Kurhaus*. All the time we meditated upon and discussed the problem of operating a tractor in the vineyard at Schloss Ramsburg. But we found no solution.

This evening I have been writing this report. And, as it is now eleven o'clock, I shall now go to bed, feeling very low in my mind, but hoping that the morrow may perhaps bring some new inspiration.

With kindest regards to all the boys—excepting the advertising department.

Most sincerely,
ALEXANDER BOTTS.

ALEXANDER BOTTS
EUROPEAN REPRESENTATIVE
FOR THE
EARTHWORM TRACTOR

MERANO, ITALY.
THURSDAY EVENING,
MARCH 28, 1928.

MR. GILBERT HENDERSON,
EARTHWORM TRACTOR COMPANY,
EARTHWORM CITY, ILLINOIS.

DEAR HENDERSON: Seven days have passed since my last letter. They have been days of terrific work and tremendous excitement. And, although we have not yet sold the tractor, we expect to do so tomorrow without fail. When you read what Gadget and I have done, and when you see the photographs, which I will send you as soon as they are developed, and which show the truly remarkable apparatus we have constructed, you will realize that we have inaugurated an entirely new era in the development of hillside vineyards.

I must begin at the beginning. You will remember that a week ago tonight Gadget and I went to bed weary and discouraged. We seemed to be licked. But after a splendid night's rest we awoke refreshed, and with a feeling of new hope. Before we had even got out of bed a great idea suddenly broke upon our minds. And it broke upon us simultaneously—such is the psychic interaction in our two minds.

"I know what we can do," said Gadget. "We can rig up an aerial cable railway like the one up the mountain over here. With that we can move the tractor to any level in the vineyard we want."

"The same thought had just occurred to me," I said. "We can buy up the abandoned railroad that is beside the new one."

"Splendid!" said Gadget. "That was my idea exactly."

"Let's go," I said.

We arose. We dressed. We grabbed a quick breakfast. We rushed over to the foot of the cable railroad. We imported the abandoned equipment. The motors and winches had been removed, and nothing remained but the cables and the two little cars, but these were all we wanted. The cables were old and rusty; they were no longer strong enough to carry heavy loads up and down a three-thousand-foot mountain, but they were obviously all right to carry a small Earthworm tractor over five hundred feet of orchard.

We rushed back to town; we called on the owner of the property, and Gadget told him that if he would pay us a thousand lire we would cart away the old cables and cars for him—thus cleaning up the landscape and making a more pleasing and artistic view for his passengers. To this he replied that he would sell us the stuff for twenty thousand lire. Whereupon we walked right out on him, and he chased after us and said he would make it ten thousand. So we went back, and after about half an hour we bought his old railroad—which was, of course, worth nothing to him—for a thousand lire, or about fifty-five dollars.

Then we put in one week of the hardest work I have ever experienced, all of which—in order not to be tedious—I will describe only in the briefest way.

We transported the materials with the tractor and a farm wagon borrowed from Schloss Ramsburg. And Gadget persuaded a bunch of the peasants at the Schloss to help us with the heavy part of the work. We ran a section of the big cable through a loophole in the twelve-foot-thick wall of the great donjon tower, and out another loophole. We carried the ends down the hill and anchored them around a jutting ledge of rock at the bottom of the vineyard; stretching them with a block and tackle hitched onto the tractor. This stretching pulled them up into the air until they were at most places forty or fifty feet above the vineyard. We mounted the two cars on these two lines of heavy cable, and connected them up with a length of the lighter cable which passed through a couple of pulleys attached to the donjon tower. This caused the two cars to balance; when one went up, the other went down. Then we fastened heavy blocks and tackles, which I borrowed from the castle stable man, to the bottom of each car. And the job was done.

As I read over the above description, it sounds very simple and easy. But it was not. When you try to manipulate over a thousand feet of heavy steel cable, you find you have a real job. The cars were harder to handle than a couple of grand pianos. And there were other problems too numerous to mention. It was indeed a titanic task which we performed, and now that it is done we want you to know that Gadget and I both feel very proud.

Tomorrow morning the Count and the Countess are due to arrive from Vienna. And tomorrow night I hope to write you that this excellent nobleman has bought the tractor and paid for the cost of installing the cable railroad.

We are now going down to the *Kurhaus* to have a cup of tea and listen to the music.

Most sincerely,
ALEXANDER BOTTS.

ALEXANDER BOTTS
EUROPEAN REPRESENTATIVE
FOR THE
EARTHWORM TRACTOR

MERANO, ITALY.
FRIDAY EVENING,
MARCH 30, 1928.

MR. GILBERT HENDERSON,
EARTHWORM TRACTOR COMPANY,
EARTHWORM CITY, ILLINOIS.

DEAR HENDERSON: The Count got back this morning, and we have demonstrated the tractor and the serial railway. And as the results have been a bit curious and unexpected, I will have to give you a complete narrative of what happened.

Gadget and I met the Count and the Countess at the station right after breakfast, and the Count was kind enough to invite us to ride up to the castle with him in his carriage. We all climbed aboard, and I wish to state that this ride gave me a feeling of aristocratic exclusiveness that would be hard to beat. The carriage was one of those open affairs known as a victoria. It was polished up like an old-fashioned barroom, and was pulled by two beautifully groomed hackney horses—the high-stepping kind. On the seat in front were a coachman and a footman—both in some sort of uniforms like naval officers. Of course, I usually prefer machinery to horses, but I will have to admit that there was an old-fashioned elegance and *je ne sais quoi* to this equipage that you could never get in a mere automobile, no matter how expensive.

When we came in sight of Schloss Ramsburg, and the Count got his first distant view of the great aerial railroad, he acted surprised and not entirely enthusiastic. Even after Gadget had attempted to explain that the railroad would solve all his problems, he still appeared doubtful, and said he was afraid it would spoil the looks of the place. Gadget very wisely refrained from arguing this point with him, merely remarking that she thought the cables added to the picturesqueness of the place, and certainly were no worse than the decorated clothesline and the manure pile in the courtyard.

To this Count Hasendorf made no reply, maintaining a thoughtful silence for the rest of the ride. The Countess, as usual, had nothing at all to say.

When we arrived at the castle I proceeded to put on a demonstration—one of the finest tractor demonstrations ever seen in this or any other country. I cranked the machine, drove it through the small gate onto the upper terrace of the vineyard, and stopped underneath Car Number One, which was suspended from its cable about forty feet above me, Car Number Two being on the other cable at the bottom of the hill. I looped a couple of logging chains around the hook of the pulley block which hung within reach. I fastened the ends of one chain to the drawbar and the ends of the other around the front part of the main frame of the tractor. Then I attached the end of the tackle rope to a winch drum which I had previously installed on the power take-off shaft on the rear of the tractor transmission case. And finally I let in the power clutch—thus setting the winch in motion—and the tractor began rising majestically into the air.

Meanwhile Gadget—who had followed along with the Count and Countess—was explaining what it was all about. "You will observe," she said, "that the little cable railway car up there is held in place by a sort of brake which grips the large cable like a vise. The power of the tractor is used only to work the block and tackle and get the machine into the air. When Mr. Botts is ready to go down the hill he will partially release the brake, which is controlled by those two small ropes you see hanging from the car. This will let him roll down along the cable until he is over whatever terrace he wishes to land on. Then he will set the brake, and let himself down to terra firma with the block and tackle. Most of the energy is thus supplied by utilizing the force of gravity. In other words, we let Sir Isaac Newton do the work. A very clever arrangement, do you not think so?"

To this question the Count made no reply.

By this time I was about thirty feet in the air, and close enough to the car so that there would be no excessive swinging about of the tractor

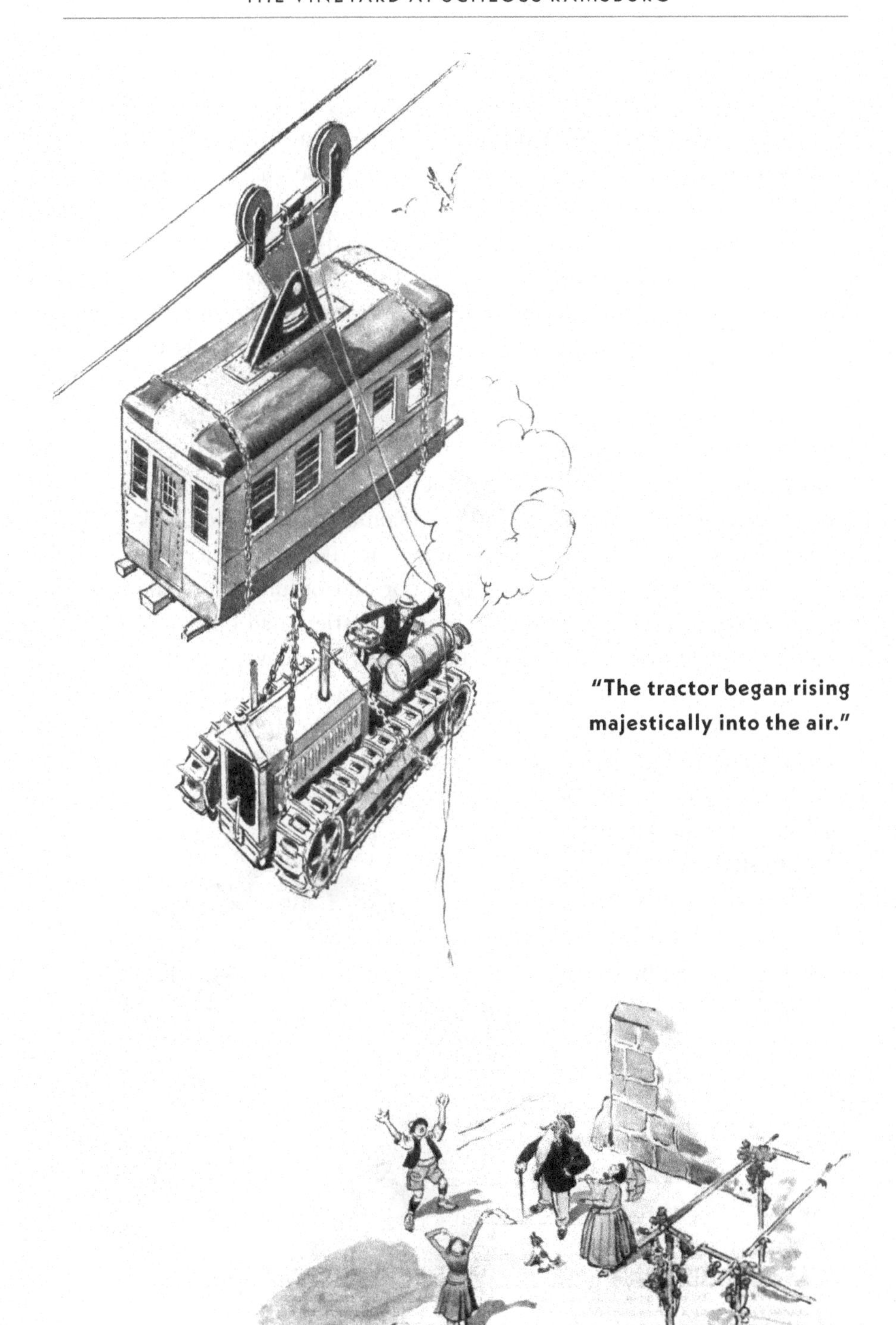

"The tractor began rising majestically into the air."

when the car moved. I shut off the power and applied the brake which was attached to the winch. So far everything had gone beautifully, but as I looked about me I will have to admit that I suddenly had a sickening feeling of doubt. I seemed to be so high up. And the big cable, which had been so heavy to move around, now looked like a mere thread from a spiderweb. Furthermore; this cable had been discarded as unsafe by the proprietors of the funicular railway. What if it should break? What if the block and tackle or the chains around the tractor should give way? I was only thirty feet above the upper terrace, but that is a long way to drop with a tractor.

And that wasn't the worst of it. The cabin went down the hill at an angle of about forty-five degrees. If the brake failed to hold, I would go coasting down that fearful slope at the speed of a cannonball. Far away and at an incredible distance below me, I could see the jagged rocky ledge where I would end up. My gaze wandered to the distant city of Merano and to the snowy peaks rising in splendor and beauty across the valley. When I had first seen this view from the castle I had been filled with admiration. But now I couldn't seem to get much enjoyment out of it.

I glanced down at the people on the upper terrace. The snowy whiskers of the Count were waving and rippling gently in the breeze. I noticed how stylish Gadget looked in her neat little coat and close-fitting hat, and I was glad she was standing a little to one side, so she wouldn't get squashed in case the tractor and I busted loose.

Then all at once it occurred to me I was getting foolish in my old age. I told myself that never before had Alexander Botts weakened in the presence of danger. He would not do so now! Resolutely, but cautiously, I pulled the brake rope. The car began to move. And slowly and smoothly it rolled down the cable until we were over the second terrace from the top. At once I set the cable brake,

"The snowy whiskers of the Count were waving and rippling gently in the breeze."

then slipped the winch brake, and descended to the ground. I threw off the chains, hooked onto a light harrow I had carried down there the day before, and dragged it up to the end of the terrace and back, cultivating the ground between the vines.

After this I took off again—this time with much greater confidence—and gracefully descended the hill. As the car I was using went down, the other car, of course, came rolling up the other cable. In the course of my descent I landed at several different levels, just to show how easy it was. Gadget and the Count and Countess followed, walking down the little flights of steps, and they caught up with me soon after I had come gently to rest beside the rocks at the extreme bottom of the vineyard. By that time a dozen or more peasants had arrived. They had been attracted by my spectacular demonstration, and were now standing around gazing at me with respectful admiration.

While I was taking the chains off the tractor, Gadget explained to the Count that if we wanted to take the machine back into the vineyard again, it would be necessary to drive it around and up to the castle by the road, hoist it up under Car Number Two—which was now at the top—and repeat the process we had just been through with Car Number One.

"Any such repetition, however," said Gadget, "would be unnecessary at this time. I feel that we have already proved to you that you can use our Earthworm tractor in your vineyard with the greatest ease."

While Gadget was speaking I was observing the Count very closely, and I began to be a little worried. Apparently he was not very enthusiastic. And when he looked at the tractor, it was with an expression of considerable distaste.

"You will have to admit," continued Gadget, "that we have solved practically all your problems. The only remaining difficulty is that your mule, Brunhilde, is still marooned up there. We are going to get busy at once, however, and rescue this excellent, but unfortunate, creature. And then you can go ahead and operate your vineyard on a completely motorized basis."

Without waiting for a reply from the Count, Gadget turned to the peasants who were standing around, and asked if they would be so kind as to assist us. They said they would, so I had them take hold of the end of the tackle rope which went through the pulley blocks on the bottom of the car, and carry it up the stone steps into the vineyard. I led the way up from one terrace to another, and they followed after, dragging the rope and thus pulling the car along. As this car was very nicely balanced against the other car, it rolled up the cable with the greatest ease. When we had reached Terrace Number Twelve, I grabbed the little brake rope and set the brake.

Then Gadget and I went into the tent and led out Brunhilde. While Gadget held the patient creature's head, I quickly strapped about her body a neat, close-fitting jacket, or brassiere, of stout canvas, which we had manufactured the day before. On the top of this garment we had fastened a couple of loops of rope; it was but the work of a moment to slip these loops through the hook on the pulley block suspended from the car. Gadget then gave the high sign to our sturdy Tyrolean peasants; they heaved on the rope, and Brunhilde was swung high in the air. I will never forget the expression which suddenly appeared on the poor animal's face—pained surprise, coupled with a look of philosophic resignation which indicated great strength of character.

When we had ascended to a height of thirty or forty feet, I eased off on the brake. As the car gently rolled down the cable, the rest of us followed down the steps, holding on to the rope so as to keep Brunhilde clear of the ground. When the car reached the bottom, we landed our passenger neatly and quietly, and I removed the canvas union suit.

"Well, Count," said Gadget, "what do you say now?"

"Wonderful! Wonderful!" he replied. He was more enthusiastic than I had believed possible. "You don't know how much you have helped me. Brunhilde is the finest mule in my stable. I had supposed she was destined to spend the rest of her life up there on that terrace. And now you have rescued her. And not only that—you have solved the problem of cultivating the vineyard."

"Well, Count," said Gadget, "we're certainly glad you are pleased with what we have done."

"I am overjoyed. What can I ever do to repay you?"

"The answer to that," said Gadget, "is easy. First of all, you can pay the somewhat modest cost of the cable railway—a thousand lire plus a few extras."

"Gladly!" said the Count. "But that is not enough. I must beg you each to accept at least another thousand for the time you have put in. Let us say three thousand lire altogether. And is there nothing else I can do?"

"There sure is. You can buy the tractor."

"Buy the tractor?"

"Yes."

"But I would have no use for it."

"No use for it? Why, you can't get along without it, count. You'll need it to cultivate your grapes."

"Oh, no. Now that you have shown me how to get mules around the vineyard, I can cultivate with them, just as I did in the good old days."

"You mean you don't want our tractor?"

"Of course not. The only reason I ever considered a tractor was because I thought there was no other way I could manage. Certainly no man in his right mind would ever put into his vineyard one of those horrible, poisonous machines if it were in any way possible to do the work as the Lord intended, by horses or mules. And that is why I am so grateful to you, my dear children."

"Oh! You feel grateful to us?"

"Believe me, I do." The Count became quite affectionate. "I sent for you as a last resort. And you, in your clever American way, have done far better than I ever expected. I had feared I would be compelled to use a tractor. You have shown me that a tractor is unnecessary. You are wonderful. I hope you will both stay for dinner."

"Thanks," said Gadget. "We will."

We stayed for dinner. It was a swell meal, with wine from the famous Schloss Ramsburg cellars. But, although Gadget talked as cleverly and persuasively as she knew how, and although the Count was most courteous and friendly, it was impossible to get him to take the slightest interest in even considering the possibility of buying such a hideous and immoral contraption as a tractor.

In the end I couldn't help but admire him. A splendid old gentleman, and an independent thinker.

He paid us three thousand lire for the cable railroad, so we made a profit on that anyway. When we finally left, he and the Countess bade us a most affectionate farewell.

"I cannot tell you," said the Count, "how surprised I was at the result of your week's work here."

"Well," said Gadget, "the result was something of a surprise to us too."

Most sincerely,
ALEXANDER BOTTS.

THE LEANING TOWER OF VENICE

ILLUSTRATED BY TONY SARG

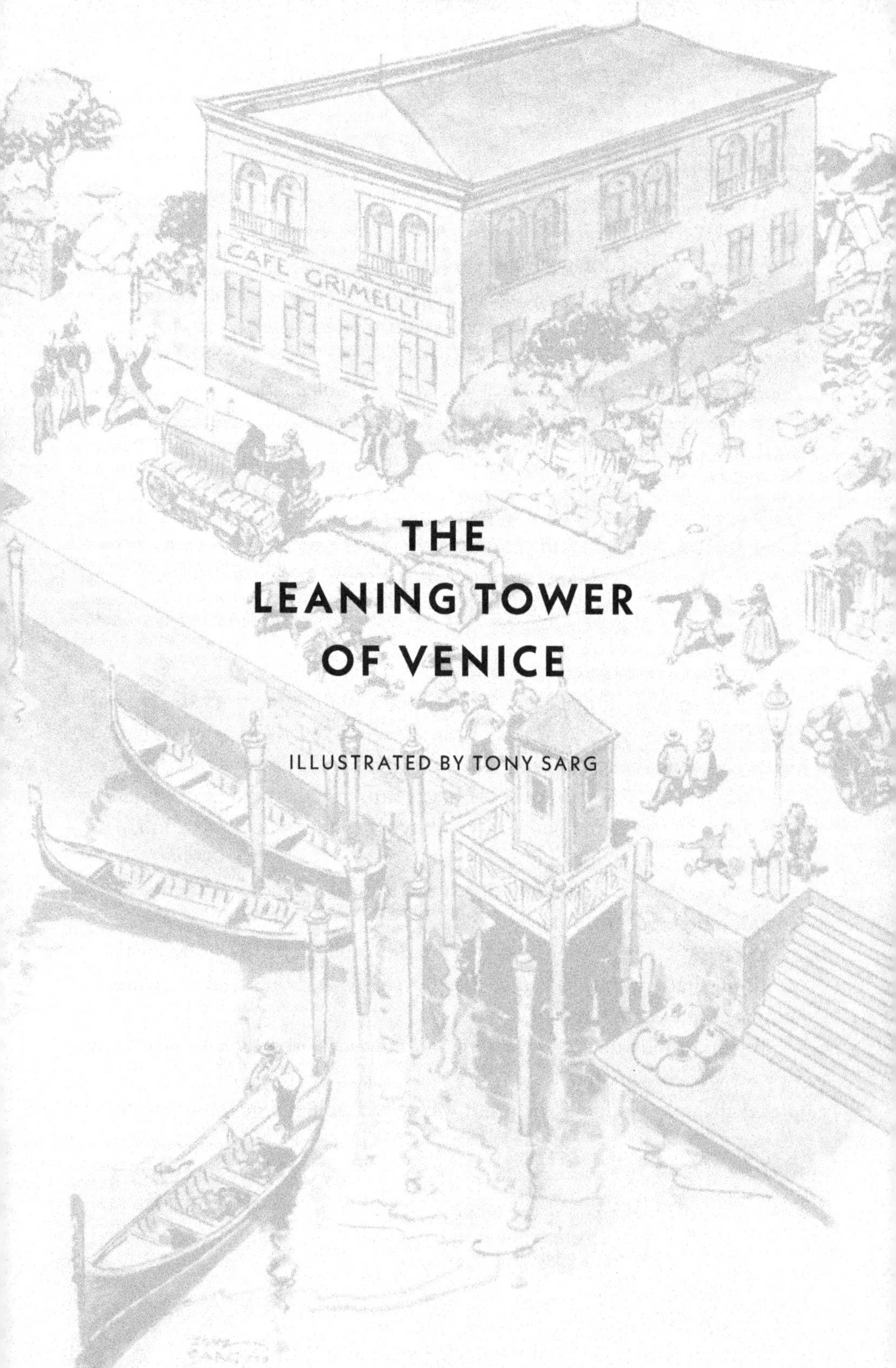

ALEXANDER BOTTS
EUROPEAN REPRESENTATIVE
FOR THE
EARTHWORM TRACTOR

MERANO, ITALY.
MONDAY, APRIL 2, 1928.

MR. GILBERT HENDERSON,
EARTHWORM TRACTOR COMPANY,
EARTHWORM CITY, ILLINOIS.

DEAR HENDERSON: Tomorrow we are leaving for Venice.

A couple of days ago I got a letter from a bozo down there who seems to be a hot prospect for a tractor. The letter was in Italian, but Gadget, of course, had no trouble in translating it. (Note: Every time I turn around over here I have to stop and congratulate myself for having the good sense to marry a woman like Gadget. She serves as wife, stenographer, interpreter of foreign languages and business adviser. And all at an expense no greater than I would have to pay for a wife alone.)

The letter from Venice ran in part as follows: "I can use one of your Earthworm tractors here at once. I have seen the advertisement which your company sent out, and I believe you have exactly the machine I need. I have a job here, and plenty of money to pay you if you can handle the work. But you will have to hurry. If you delay even a few days, it may be too late. Please come at once, and bring your tractor with you."

This letter was signed by a Signor Luigi Bontade. He does not say who he is, nor does he describe his business or the particular work for which he needs a tractor. At first I was not inclined to take the matter very seriously.

"Obviously," I said to Gadget, "the man is crazy. How could he use a tractor in Venice? I have seen pictures of the place, and all the streets are canals. Does he think our machine can walk on the water?"

"Possibly he does," said Gadget. "We have run into people who had even funnier ideas than that."

"It is probably that idiotic advertising department again," I suggested. "I could even make a guess at exactly what they have done. Do you remember that booklet they put out called 'The Earthworm in the Rice Fields'?"

"Yes," said Gadget. "It had pictures of tractors going through the water."

"Exactly," I said. "They were going through water about six inches deep—but you can't tell that from the picture. And when these advertising experts get some poor, ignorant translator to turn their text into Italian, it may mean anything at all. You remember how one of their linguists used the French word '*rochers*' to translate Rocky Mountains, and made it appear as if the tractor could climb right up a vertical cliff."

"I remember that very well," said Gadget. "How could I ever forget it?"

"And they have probably made a similar boner down here. They have made it look as if the tractor was ambidextrous, or amphibious, or whatever you call it. And this guy thinks he can use it like a boat."

"That might be so," said Gadget. "But again it might not. They have canals in Venice, but they also have streets."

"Where did you get that idea?"

"People who have been there have told me. There are several open paved squares, and there is a network of little streets that are used by pedestrians only."

"They don't allow wagons or automobiles?"

"No."

"If they don't allow wagons and automobiles, they wouldn't allow tractors, would they?"

"Maybe not. But there might be other places they could use them. They have glass factories in Venice."

"Yeah?"

"And they might use a tractor in one of those to haul sand or cinders or something."

"It doesn't sound promising," I said. "I think we had better write to Signor Luigi Bontade and ask him for particulars."

"If we did that," said Gadget, "it might delay matters too long. He says he is in a great hurry. He needs the tractor at once."

"Well," I said, "we have no other leads right now that are any better. I suppose we might as well take a chance and go down and see what it is all about."

"Exactly," said Gadget. "And besides, we have never been to Venice, and I think we owe it to ourselves to have a look at it. From all accounts, it must be quite a place."

"Right you are," I said. "We will go." And so it was decided.

We have already shipped the tractor by express, or *grande vitesse*, or something, which the guy says is much faster than ordinary freight. And we are leaving ourselves tomorrow. I will let you know how we come out. I am not very optimistic. However, if it is possible to sell a tractor in Venice, I am the guy that can do it.

Sincerely,
ALEXANDER BOTTS.

———

ALEXANDER BOTTS
EUROPEAN REPRESENTATIVE
FOR THE
EARTHWORM TRACTOR

VENICE, ITALY.
TUESDAY, APRIL 3, 1928, 11 P.M.

MR. GILBERT HENDERSON,
EARTHWORM TRACTOR COMPANY,
EARTHWORM CITY, ILLINOIS.

DEAR HENDERSON: Gadget and I arrived in this beautiful city late this afternoon. We had telegraphed ahead to Signor Luigi Bontade, and he met us at the station. He does not want to use the tractor in a glass factory. Neither does he want to use it as a boat for cruising around the canals. Apparently the advertising department, by some accident, did a reasonably creditable job on their Italian booklet, and succeeded in giving a fairly good idea of what the Earthworm tractor is and what it can do. But that does not help in this case, because Signor Luigi Bontade—who, by the way, is a restaurant keeper—has figured up out of his rich Italian imagination a tractor job which is entirely new and unprecedented, and so wild and woolly that I don't want anything to do with it.

Signor Bontade is a large man, perhaps fifty years of age, with black hair, black mustaches, and a highly nervous temperament. When he met us at the railroad station, he greeted us with such a cordial smile that we both decided we were going to like him. He speaks English after a fashion, and he addressed us something like this:

"Ah, Signor Botts, I am delight! Ah, Signora Botts, I am charm! You have notta forget da granda machine? You have bring da beegs tract?"

Note: This guy really does talk English just about like that. But as it is too much of a job to write it down that way I will hereafter give you his remarks in ordinary English.

"I wish to use the tractor this very night," he explained. "Has it arrived?"

"It should be here," I replied. "If you will take us around to the express office we will find out for sure."

Signor Luigi led us through the station to a small bureau with a little window like a ticket office. The man in the window informed us that the tractor had arrived, but that we could not get it until tomorrow. This remark threw our friend into a state of tremendous excitement. He poured forth a surprising stream of language; he danced about; he waved his arms; he became red in the face, and then almost purple. But the express agent remained stubborn. It appeared that they were just closing up the office for the night and they would not be open again until eight in the morning. That was his story, and he stuck to it, finally ending the argument by closing the shutter of the little window.

"This remark threw our friend into a state of tremendous excitement. He poured forth a surprising stream of language."

"Oh, well," I said, "tomorrow morning will be all right. I don't know what you want to use the machine for, but whatever it is, we can probably do it better by daylight."

"No, no," said Luigi. "I want it tonight."

"Apparently you can't have it," I said. "So you might as well make up your mind to wait until tomorrow. And in the meantime you might tell us for what purpose you wish to buy this tractor."

"I don't want to buy it. I just want to rent it."

"I knew there would be some catch to it," I said sadly. "How long do you figure on renting this machine?"

"Just for one night."

"What? Do you mean to say that you got us to come all this distance with a tractor which we thought you wanted to buy and which you now admit you only want to rent for a single day?"

"Not for a day," Luigi said. "For a night. This job must be done during the hours of darkness. But it will be all right. I will pay you very well. I will pay you most lavishly."

"My business," I said, "is selling tractors, not renting them. If I had known you did not want to buy, I would not have come at all."

"You will not lose anything," said Signor Bontade. "I will pay you very well for your time and for the use of the tractor."

"Well," I said, "as long as we are here we might as well investigate your proposition anyway. Please explain to us what you want to do."

"I will. I will take you to my restaurant so you can see for yourself exactly how the land lies and exactly what is to be done. Come with me."

We followed him through the station and onto the Fondamenta Santa Lucia, which runs along the Grand Canal. Luigi engaged one of the shouting mob of gondoliers, who took our suitcases and put them into his gondola. Then we all got aboard and started off gliding gently under the Ponte Alla Stazione and on toward the center of the city.

And right here I might as well admit that everything these poets have written about Venice is true. Anytime that you guys in Earthworm City, Illinois, get to feeling proud of the town and your old courthouse, your Masonic Temple and your brand-new moving-picture theater, you had better come over here and get an eyeful of something that really amounts to something.

As we floated along through the twilight, Gadget and I were both so impressed we couldn't say a word. The canal is lined on both sides by ancient palaces which are built of stone and carved in the most delicate and interesting way. The stone—which I think is mostly marble—is old and weathered. It is of various colors—creamy white, misty bluish gray and delicate rose. And when these tints are reflected in the water they become even more misty and beautiful than ever. A few yellow lights were beginning to show in windows, and we passed several boats with lanterns. We could hear the distant shouts of the gondoliers, the lapping of little waves against the sides of the canal, and the creak and dipping of the oars. The lack of other noises, after the hubbub in the station, was almost startling.

But unfortunately the silence did not last. Luigi, although a good egg and a likable fellow, did not seem to be impressed by the dreamy beauty of the scene.

"That pig of an express agent," he announced in a loud voice, "makes me sick. He has delayed us twenty-four hours; and for no reason at all except that he is too lazy to do a little work after closing time. It is an outrage."

"Well," said Gadget, "apparently we can't help it. We might as well stop worrying about it."

"You are right, *signora*. You are right. One should be philosophical. But it is hard when troubles come so thick, so fast, so heavy."

"You have been having hard luck?" asked Gadget.

"Hard luck!" said Lulgi. "It has been insupportable. Never in the entire history of the world, *signora*, has there been a man so overwhelmed with trouble, disaster and tragedy."

"I am sorry," said Gadget. "Just where does it seem to hurt you most?"

"I will tell you," he said. "I am a restaurant keeper. I own a restaurant."

"I don't see anything so tragic about that."

"But wait, *signora*, wait until I tell you all."

"Go on," said Gadget.

"My restaurant is the Café Bontade. It is not one of the large places on the Piazza San Marco. It is a modest establishment on the Grand Canal, not far from the Rialto."

Note: As most of you bozos at the home office have not had the educational advantages of European travel, I will explain that a piazza in not a back porch, or even a front porch. It is a public square. The Piazza San Marco is the most important and fashionable square in Venice. And the Rialto is a bridge—not a moving-picture show.

"Although my place is small," continued Luigi, "I have always had a very good tourist trade. The chief attraction is my beautiful little garden, which overlooks the Grand Canal. Here, beneath the shade of the trees, I have tables and chairs where many customers have come to sit at their ease, eating our excellent food, sipping our excellent wines, listening to the sweet singing of the birds, and watching the gondolas gliding by. I have run the Café Bontade for ten years. It has been profitable; I have made money."

"Well," said Gadget, "I do not see any cause for shedding tears over that."

"Wait," said Luigi. "I am now coming to the dreadful part of it. I have a competitor. Next door to me is another restaurant. It is called the Café Grimelli, and it is owned and operated by a pig of a man called Antonio Grimelli."

"And he has been cutting in on your trade?" asked Gadget.

"For ten years he has been trying to get my customers away from me. He has fitted up a little garden in imitation of mine. He has reduced his prices. He has put out a large sign. He has bribed many of the gondoliers to bring their passengers to his place, even when they have especially commanded that they be taken to mine. But all these things have availed him nothing, because even the tourists know the difference between good food and bad."

"You mean his stuff is not so good as yours?"

"There is no comparison, *signora*. His food, his wine, his service—everything is unspeakable. As a citizen of Venice I am ashamed that our fair city should be polluted with the presence of such an establishment. His chefs are worse than the so-called cooks in the army."

"They must be pretty bad," I said.

"They are," said Luigi. "The soups in the Café Grimelli would not be fit to use as slop for hogs; the hors d'oeuvres are mere aggregations of garbage; the pièces de résistance are tough, tasteless and sickening; the wines and the desserts are beneath contempt. And that is not all."

"No?" said Gadget.

"This wretched material that he calls food," continued Luigi, "is prepared in kitchens that swarm with rats and cockroaches. It is carried to the tables by frowzy waiters with black fingernails and shoulders that are powdered with dandruff."

"And I suppose," I said, "they all have halitosis?"

"I do not know what that is. But if it is something disagreeable, they probably have it. The spoons they use are always greasy."

"In other words," I said, "this Antonio Grimelli does not run what might be called a first-class restaurant. But if this is so, I don't see why you should fear his competition."

"I would not fear Antonio if he competed with me in an honest and fair manner. But recently he has been using infamous and dishonorable tactics."

"What has he done?"

"You would hardly believe it, but this criminal owns a big tower."

"I don't see anything criminal about that. Has he just recently bought it?"

"No, no. It is an old structure built many hundred years ago. It stands on Antonio Grimelli's property, just across the boundary line from my restaurant garden."

"If it is on his property," said Gadget, "I don't see why it should bother you."

"It is on his property now," said Luigi. "But I fear that it may soon be on mine."

"You mean this Antonio is purporting to move his tower?"

"No, he is not."

"Well, just what is he doing then?"

"He is doing nothing."

"Nothing?"

"Yes, that is just the trouble. I have pleaded with him; he only laughs at me. I have appealed to his sense of justice; apparently he has none. Finally, I have brought suit against him in the courts of Venice, but he only snaps his fingers in my face. He has hired clever lawyers who have got the matter so tied up with adjournments and delays that my own lawyer tells me I can hope for no relief for weeks or months to come."

"I don't understand," said Gadget, "why you should get excited because this guy owns a tower and doesn't do anything with it?"

"It is because the structure is out of plumb, *signora*. It is call the *Torre Pendente*, which means 'hanging or leaning tower.'"

"Something like the famous Leaning Tower of Pisa?"

"Exactly," said Luigi. "But there is this difference. The tower at Pisa can lean all it wants to, and it doesn't bother me at all. But this Torre Pendente here leans right out over the garden where I have all my restaurant business. And I don't like it."

"How long has this thing been a leaning tower?"

"As long as anyone can remember, *signora*. It has always been called the Torre Pendente."

"Then why, if it has always been leaning the same way, should you get in a panic about it all at once?"

"When I started my restaurant ten years ago," said Luigi, "the top of the tower was leaning nearly a foot over toward my property. It had been in this position for many years, and it remained the same until this winter. It seemed perfectly strong and solid. But suddenly, one day two weeks ago, it was moved."

"This Antonio Grimelli moved it?"

"No, it moved itself."

"How could it do that?"

"The lower side of it settled."

"Why did it settle?"

"You must understand," said Luigi, "that this beautiful golden city of Venice has feet of clay. It is not a city founded on a rock; it is built on the mud and sand of the lagoon. The men of olden times, when Venice was the queen of the world, were compelled—in order to make their buildings stand up at all—to drive down wooden piles. These piles made a solid foundation for the time being. But now the wood is rotting away and it seems as though our wonderful city were crumbling. It is sad—very sad."

I looked around apprehensively. The reflections of the palaces along the canal were trembling and wabbling in the rippling surface of the water, but the buildings themselves looked solid and secure.

"You need not be afraid," said Luigi. "The whole city will not fall at once. It is a slow process, but it seems inevitable. When you visit the Basilica di San Marco you will see that the arches are cracking. The mosaic floor has sagged in many places. When you go to visit the Campanile di San Marco, you will see that it is all new construction. The original Campanile fell down into the Piazza on July 14, 1902."

"And you are afraid that the tower next door to your restaurant will fall down in the same way?"

"I am. It was just two weeks ago that the foundation partially collapsed. I will never forget that afternoon. It was warm. The sun was shining pleasantly. The garden was full of customers. There came a faint, low, rumbling noise. The ground appeared to tremble ever so slightly. Dishes and spoons rattled on the tables. We all looked up. The tower was moving! Slowly and very majestically that mass of masonry was tilting itself out over the garden."

"It must have given you quite a scare," said Gadget.

"It did, *signora*. I shouted a warning. Others took up the cry. 'Run for your, lives!' they yelled. 'The tower is falling!' And with wild shrieks all my customers stampeded out of the garden. They upset the tables. They broke many dishes. Several of them fell down and were stepped on—but not seriously injured, thank the Lord. After we had all got a safe distance away and the noise and excitement of the panic had somewhat subsided, we heard pitiful cries for help from the canal, and we were just in time to fish out an elderly British colonel who, in the excitement of the moment, had fallen into the water. It was a terrible occasion."

"It must have been," said Gadget. "But you say the tower did not fall after all?"

"It did not," replied Luigi. "Or at least it hasn't yet. The top of the tower moved about two feet, and then stopped. But my restaurant business is ruined as completely as if the tower had actually come crashing down. It was two weeks ago that my customers were scared out by that awe-inspiring event, and not one of them has returned. No one with any judgment would take the risk of eating in that garden now. And even if anyone were as foolhardy, he would not be permitted to. The *vigili*—the police—have roped off the entire area and will allow no one to enter."

"Haven't you got any other place they could eat?" asked Gadget.

"We have an indoor dining room which we use in cool or rainy weather. But the garden has always been the chief attraction of our establishment. Besides, we cannot use the dining room now anyway."

"Why not?"

"Because the entrance to it is through the garden."

"That is the only entrance you have?"

"You can come around by way of a little back alley, into a rear door, down a long dark hallway, through the kitchen, and from there into the dining room. That is the only entrance we now have for ourselves."

"That certainly is tough luck," said Gadget. "And I suppose you cannot expect the tourists to come in through such a roundabout and difficult entrance."

"Exactly. And now that the pleasant spring weather is here, they don't want to eat inside anyway. They want to eat outdoors. And they all go to Antonio Grimelli's garden. They do not know what a vile hole it is until they have started to eat. And they are so disgusted then that they do not want to eat anywhere. But here we are."

The gondola stopped at the foot of some steps which led up to the top of the stone embankment along the canal. We disembarked.

Note: The above remarks of Signor Luigi Bontade were spoken, as I have said, in a most atrocious brand of English. But as I did not wish either to bother myself by reporting it exactly, or to fatigue you people by forcing you to read it, I have taken the liberty of transcribing it into the polished and beautiful English which you have just read.

Luigi insisted on carrying our suitcases. When we reached the top of the steps he set them down. He placed two fingers in his mouth and whistled loudly. Soon a man appeared and took charge of our baggage.

Then Luigi proceeded to point out the various features of the landscape. We were standing on a narrow walk or promenade which followed the bank of the canal. In front of us was a handsome stone building bearing a sign—CAFÉ BONTADE. This was Luigi's restaurant. To the left of it was his garden—a beautiful little place with many small trees and flowering shrubs. It was roped off, and there were warning signs which had been posted by the police. On the far side of the garden was a stone building with the sign, CAFÉ GRIMELLI, and beyond this another garden with many tables at which people were eating. Evidently the obnoxious Antonio Grimelli was doing a good business.

At the rear of Antonio's building was the famous Torre Pendente. It was about fifty feet high, I should say, built of creamy white stone that must have been marble, and carved and ornamented in a very lovely fashion. It had balconies with intricately carved stone railings, and windows with ornate Moorish arches and little twisted pillars. But the most remarkable thing about it was the angle at which it stood. It leaned so far out over Luigi's garden that it looked as if a mere puff of wind, or possibly even a robin alighting on top of it, would be enough to bring it down.

"Your restaurant business," I said, "certainly looks to be in a bad way. It seems to me it is up to Antonio to do something about that tower."

"That is what I say," said Luigi.

"He ought to have it straightened up and reinforced," I said, "or else he ought to have it torn down."

"He ought to, but he won't. And it is easy to see why. As long as this tower hangs threateningly above the garden, my business is ruined, while he grows rich from the customers who normally would come to me."

"I should think," I said, "he would fix it of his own free will. If it falls down he will lose a very handsome tower."

"He is taking a chance. He has had it examined by an engineer who thinks it has settled all it is going to for the time being. The engineer is not certain, of course, but he thinks it will be safe for several years at least."

"If that is the case, why can't you go ahead and use your garden?"

"The engineer is not absolutely sure. I do not wish to endanger my customers. And even if I did, the police would not let me."

"And you say you have brought a court action against this Antonio?"

"I have. But as I told you, he has the matter so tied up with technicalities and lawyers' quibblings that it will be months before I can accomplish anything."

"And what are you going to do in the meantime?" I asked.

Luigi smiled a grim and self-satisfied smile. I could see his white teeth glistening through the gathering twilight.

"I am going to take matters into my own hands," he said. "You and I and your excellent American Earthworm tractor are going to fix this Antonio Grimelli."

"Oh!" I said, somewhat startled. "So that is the idea?"

"That is the idea," he said, still grinning through the darkness.

"And just what," I asked, "did you have in mind? In what way do you plan to use the tractors?"

"It is very simple," said Luigi. "If Antonio will not straighten up his tower and keep it on his own property, where it belongs, we will do it for him."

"But that is no job for a tractor. What you need is a lot of jacks to jack up the low side of the tower. Then the foundation should be reinforced. I have never had any experience in this kind of work, but I am almost sure there are methods of inserting a tube under a weak foundation and forcing in concrete."

"That would take too long." said Luigi. "As soon as Antonio found out what we were doing he would have us thrown off his property as trespassers. No, we must use a tractor, which will do the work quickly."

"I am afraid that I don't understand you even yet."

"Come," said Luigi. "I will show you what I have in mind."

He led us along the walk which bordered the Grand Canal. We passed in front of his garden, in front of Antonio's building, and in front of Antonio's garden. Beyond this garden, separated from it by a low stone wall, was a small public square which they called a *campo*. Luigi took us halfway across the campo and then stopped. He motioned for Gadget and me to come very close to him, and when we had done so, he began speaking in a hoarse and mysterious whisper.

"Tomorrow," he said, "we will get the tractor and load it onto a barge. We will cover it with a canvas so no one will know what it is. I will borrow a long cable from one of my friends who is in the ship-supply business. We will put the cable on the barge. I will also procure a ladder."

"You don't have to whisper," said Gadget. "There isn't anybody anywhere near."

Luigi glanced furtively about. "It is always well to be careful," he said. Then he continued in a whisper that was more mysterious than ever: "We will wait at the station until after midnight. Then, in the very darkest hour before dawn, we will row away from the station and proceed stealthily along the Grand Canal."

"Why stealthily?" asked Gadget.

"So that no one will see us."

"And why don't we want anyone to see us?"

"Hush! Not so loud! If anyone saw us, it might spoil all our plans. When I have explained everything you will understand."

"All right. Go on."

"We will row the barge along the canal until we arrive here at this campo. We will tie up the boat, throw across our gangplank, and get everything in readiness to run the tractor ashore. Then we will leave it."

"You mean we won't unload the tractor?" asked Gadget.

"Not at once. That will come later. First of all we will take the cable and the ladder, and carry them over here."

He grasped us by the arms and led us away from the canal to a point about in the middle of the open campo. He then pointed to the leaning tower, which was visible over the low wall of Antonio's garden. By this time the daylight was almost entirely gone. But there was plenty of artificial light from the street lamps and from the lanterns in the garden. From where we stood, the tower looked perfectly straight. It was, in fact, leaning directly away from us.

"And now," continued Luigi, "I come to the really brilliant part of my plan. When we arrive here tomorrow night, it will, of course, be dark. The great city of Venice will be asleep, and the campo will be deserted. I have made discreet inquiries and I find that in the early morning hours the nearest policeman is a quarter of a mile away."

"You seem to have planned this very elaborately," said Gadget.

"I have, *signora*." Luigi smiled, and once more I could see his teeth gleaming through the dusk. "I will take the end of the cable," he continued, "and carry it over the wall and across the rear part of Antonio's garden. Then, by means of the ladder, I will climb up to the roof of Antonio's house. I will haul the end of the cable and the ladder up after me. Placing the bottom of the ladder on the roof, I will lean the top of the ladder against the tower. Still carrying the end of the cable, I will climb up to the little balcony which encircles the tower near the top. I will walk around the balcony, dragging the cable after me, and make a loop inclosing the tower. I will then descend to the ground and rejoin you here in the campo. Have I made myself clear? Do you visualize the scene as it will then exist?"

"Yes," said Gadget. "One end of the cable will be fastened around the upper part of the tower; the other end will be down here."

"You are right" said Luigi. "And what do you think we are going to do next?"

"This seems to be your story," said Gadget. "Suppose you go on and finish it."

"I will," said Luigi. "It is now that Signor Botts becomes active. He will crank up his tractor, drive it off the barge, and bring it up here to the middle of the campo. He will hook onto the end of the cable. He will cause his machine to exert all the power that is in it, and he will give that tower a terrific pull. Then we will rush the tractor back onto the barge, and under cover of the night we will glide stealthily back to the station."

"Wait a minute," I said. "All this may sound OK to you, but it listens pretty queer to me. Where did you get the idea that we could do anything like this?"

"I got the idea," said Luigi, proudly, "from the tractor advertisement sent out by your company."

"Oh, yeah?" I said.

"There was a picture showing one of your tractors pulling down a tremendous factory chimney. For some reason or other they wished to demolish the chimney; and they attached a long cable to it and pulled it down with a tractor."

"That part of it is all right," I said. "I remember the advertisement and I remember the picture."

"If your tractor could handle a big chimney like that, it could certainly handle a medium-sized tower like this."

"But this is a different proposition," I said. "As I understand it you want to pull this tower up straight, so that it won't threaten your garden any more. It is possible, of course, that we might be able to do this; although it would be a very risky business. But even if we could, it wouldn't help us any, because as soon as we slacked off on the cable, the tower would settle back to where it was before. It might even go back farther."

"Who said anything about slacking off?" said Luigi. "My idea is to pull it up straight and then keep right on going and drag it over until it falls down into Antonio's garden."

"You are crazy," I said. "You ought to know you can't get by with anything like that."

"Why not? The place will be deserted. There will be nobody to stop us."

"But what will they do to us afterwards?"

"Nothing," said Luigi. "We will get away so quick that they won't know how it happened. They'll think it fell down naturally."

"But suppose somebody was in the tower?"

"The tower is usually empty at night. If Antonio happened to be in there, it would just be his hard luck. And it would serve him right for being so nasty to me."

"Well, honestly, Luigi," I said, "this scheme of yours is the most remarkable I have ever run across."

"Then you approve?" he said. "I am so glad. Tomorrow night we will go ahead and put my plan into operation. It is all settled."

"Settled nothing!" I said. "I most certainly will not have anything to do with this business at all."

"But I thought you said you liked my scheme?"

"I said it was remarkable. And it is; it is so remarkable that it is entirely out of my class."

Luigi seemed very much hurt. "Then you do not want to help me?" he asked, in an incredulous voice.

"I should like nothing better than to help you," I replied, "but not in any such way as you have described."

"What other way is there?"

"I don't know."

"But I must do something," he said. "I am in a desperate predicament. My business is being completely ruined."

"I am very sorry indeed for your troubles. I admit that you are perfectly justified when you say you are in a desperate predicament. But I am not; and I don't intend to get into one. I will be perfectly frank with you. Mrs. Botts and I are both peaceful, inoffensive, law-abiding people. At the moment we happen to be out of jail, and we prefer to remain so. . . . Do you not agree with me, Gadget?"

"Absolutely," said Gadget. "I don't want to be in jail anywhere. And especially not in Venice. I have a feeling that the dungeons in this burg would have a tendency to be a bit damp."

"Then you will not help me with my plan?" asked Luigi.

"No," I said. "No, no, no. And again, no."

"You will not even consider it?"

"No," I said.

Luigi looked at Gadget.

"No," said Gadget.

"Listen," said Luigi. "Perhaps I have been too hasty in asking you for your answer. Suppose you think it over tonight. And tomorrow you can give me your answer."

"It will still be just the same," I said.

"Never mind about that now," he said. "All I ask is that you spend the night with me as my guests, that you think this matter over, and that you discuss it with me once more in the morning."

"It wouldn't make any difference how long I thought it over," I said. "My answer would still be no."

"But at least you will spend the night at my house? It will cost you nothing."

"Thanks," I said. "We had really expected to stay at a hotel. We don't want to sponge on you—especially as we don't seem to be doing anything to help you out of your troubles."

"You will not impose on me at all," said Luigi. "And you will be putting yourself under no obligation. It was I who invited you to Venice. If you refuse to be my guests I will be deeply insulted."

"In that case," I said, "we will stay. And we thank you very much. But we absolutely refuse to have anything to do with any outlaw wrecking parties. So don't get your hopes up too high."

Luigi led us back to his restaurant where he introduced us to his wife. She was a pleasant, motherly looking soul, who spoke no English at all.

We had a splendid meal in the deserted dining room. Certainly Luigi was no liar when he told us he served good food in his joint.

After dinner he showed us up to our room, where I have been spending the evening writing this report. I am really getting very much interested in Luigi and his troubles. He is a very nice old bozo, in spite of his wild ideas, and I find that I am getting to like him a lot. But I can't see for the life of me how we can help him much. The only scheme that I can think of at the moment is the possibility of propping up the tower with a lot of timbers. If we did this the timbers would all be on Luigi's property, and Antonio could not interfere with them in any way. But I feel this idea doesn't amount to much. In the first place, it would not help us to sell a tractor. And in the second place, it would fill up Luigi's garden with lumber to such an extent that there wouldn't be much room left for the cash customers.

Do not think, however, that I have completely given up hope. Gadget and I will sleep on this problem tonight, and tomorrow we may get an inspiration. If we hit on any good ideas, or if anything new comes up, I will let you know.

Very sincerely,
ALEXANDER BOTTS.

ALEXANDER BOTTS
EUROPEAN REPRESENTATIVE
FOR THE
EARTHWORM TRACTOR

VENICE, ITALY.
WEDNESDAY, APRIL 4, 1928.

MR. GILBERT HENDERSON,
EARTHWORM TRACTOR COMPANY,
EARTHWORM CITY, ILLINOIS.

DEAR HENDERSON: I promised to write you as soon as anything came up, but the first thing I have to report is that something has come down. At exactly 7:32 this morning just as we were finishing breakfast in the Bontade dining room, we were subjected to a mild earthquake, accompanied by a tremendous crash outside. We rushed to the window and the first thing we saw was a cloud of dust in the garden.

As this dust cleared away it revealed a large and confused heap of building stone. The good old leaning tower had leaned a little too far and finally let go. And great was the fall thereof.

But even greater was the excitement thereafter. Hundreds of people came streaming in from all directions. The few cops that arrived had all they could do to hold back the crowds, and to keep them from pushing one another into the canal. Meanwhile the canal itself became blocked with a great traffic jam of gondolas and boats, whose occupants were craning their necks to see what was going on. Luigi and his wife began rushing hither and thither, uttering strange noises, waving their hands about, and acting exactly like a couple of highly emotional and excitable Italians—which is just what they are. Gadget and I, of course, remained cool and collected. It took me but a moment to size up the situation and to decide upon a plan of action.

I grabbed hold of Luigi and forced him to stop running around in circles and listen to me. "The first thing to be done," I said, "is to clean all this rubbish out of your garden."

"Yes, yes," he said, wildly. "But how? There are tons and tons of that rock. Even with a large force of workmen, it would take weeks, probably months. And in the meantime my business will be ruined. What can we do?"

"Listen," I said. "Answer me just one question. Do you want your garden cleaned up in a hurry?"

"Of course I do."

"Fine," I said. "It shall be done."

Without even waiting to get my hat, I pushed my way through the crowd and got hold of one of these floating Venetian taxicabs. I proceeded to the station, and after a certain amount of delay I got the tractor loaded onto a barge and conveyed to the Café Bontade. By this time the crowd, after finding out that nobody was killed, had dispersed to some extent. And as soon as I had unloaded the tractor I drove it into Luigi's garden. Fortunately, I had with me a logging chain which I had inadvertently carried away in the tractor from Schloss Ramsburg, after our attempt to make a sale to Count Hasendorf. I asked Luigi for someone to help me, and he gave me the porter of his establishment—a strong, intelligent young man by the name of Marco. I was pleased to find that Marco could speak English. I had him put the chain around a large block of marble near the edge of the heap; I hooked on with the tractor and skidded it out of the garden onto the paved promenade that runs along the canal.

"Luigi," I yelled, "where shall I take this stuff?"

"I don't know," he said.

"It belongs to your friend Antonio. How would it be if I took it over and delivered it to him in his garden?"

Luigi smiled. Then he laughed. "A splendid idea," he said. "We will turn the tables on this loathsome snake of an Antonio. If he can dump his marble into my garden, we can take it back and dump it into his garden."

"Right you are," I said. I threw in the clutch, stepped on the gas, and dragged the big hunk of stone down the promenade, in through the open gate of Antonio's garden, and back to the rear wall. After a wide sweep across a couple of flower beds, I cast off the chain and speeded back to Luigi's garden before Antonio had time to realize what was going on.

I then made three more rapid and successful trips, snaking large marble blocks after me. But on the fifth trip I found Antonio's gate barred by Antonio himself, and a couple of cops.

This was the first time I had met Signor Antonio. He was a large man with a face that may well be described as swarthy, ugly and repulsive. I did not like his looks. The two cops appeared to be good guys, but there was an air of determination about them that was not reassuring. I stopped. Luigi and young Marco came up and there was a long argument. Finally Luigi told me that it was no use. Antonio had appealed to the police to protect his garden. The police had decided it was a reasonable request, and they ordered us to keep out.

"All right," I said. "'If he won't accept delivery on his stuff, we will get rid of it somewhere else."

I drove back, hooked onto another hunk of stone, and snaked it to the parapet, where Marco, with a bar which he had found somewhere, eased it over into the canal. It disappeared with a tremendous splash. I repeated the process several times But before long a couple of other cops arrived. Luigi came running up and there was another long argument. At length Luigi turned to me.

"They won't let us do this either. It's against the law to fill up the canal this way."

"All right," I said. "Then I suppose we will have to hire some of these freight gondolas or barges, or whatever you call them, and haul the stuff away somewhere."

"That is a good idea," said Luigi. "I will get a barge right away, and we will take the stuff out into the lagoon and dump it."

He rushed off, and while he was gone I hauled a load of stone out to the edge of the canal. Before long he came riding back in a battered old boat which he had hired. At once we started putting the stone aboard.

"They certainly can't stop us from doing this," said Luigi. "You just keep on with the work. I am going to have a short conference with my lawyer. I am going to have him draw up a claim for damages that will just about ruin that louse of an Antonio. I have prevailed upon my wife to go to bed. We will claim that she has suffered a nervous breakdown from this occurrence. We will ask fifty thousand lire in damages for that. Then there will be another fifty thousand for damages to my business and to my property. Goodbye. I will see you later."

Luigi got into a gondola and disappeared around the bend of the canal. Immediately thereafter Antonio came out and tried to make us stop our work. But as we paid no attention to him, and as he could not seem to get the cops interested in helping him, he finally went away.

We finished loading the barge, and it sailed away. Then we started skidding another load of stone out to the edge of the canal.

While we were engaged in this work I was surprised to see Gadget arriving over the water of the canal in a gondola. It suddenly occurred to me that she had been gone all morning. I had been so busy that I had hardly noticed her absence. When she disembarked I noticed that she was accompanied by a very nice-looking middle-aged Italian gentleman. This gentleman she introduced as Signor Sagredo. He had heard of the fall of the tower, Gadget explained, and was very anxious to inspect the ruins.

"The ruins," I said, "are easy to inspect. We shall be glad to have you walk around and look at them as much as you want. And in the meantime you will excuse me if I go back to work."

Marco and I returned to our task. By this time my young helper had become very much interested in the tractor. He told me he had been a mechanic in the army; he understood machinery, and wanted to learn to drive the tractor. This seemed to me like a very good idea. I let him get into the driver's seat, and after a little instruction and practice he was able to drive almost as well as I could myself.

In the meantime Gadget and her boyfriend were walking about examining the remains of the tower, and carrying on a very earnest discussion in Italian. At length he said goodbye and departed in a gondola. Soon after, Luigi returned from his visit to this lawyer. By this time I had accumulated enough stone along the side of the canal for another boatload.

"You have done very well," said Luigi. "It is now after twelve o'clock; it is time we had something to eat."

"Very good," I said.

Marco ran the tractor into the garden and shut off the motor. We all went into the house. And as we sat down to one of Luigi's most excellent repasts, I decided that the time was ripe to talk a little business.

"From what we have done so far," I said, "I calculate that it will take ten days to clean all of the debris out of your garden. To keep up to schedule, however, we will have to have at least two boats to haul the stuff away."

"I will rent another boat this afternoon," said Luigi. "And I might as well tell you that I am very much impressed with your tractor. I have been watching it work, and I am perfectly amazed at the speed with which it moves those big stones. It looks to me as if you could finish the job in less than ten days, whereas, if I tried to do it by hand, it would take several months."

"You have decided then," I said, "that you want the tractor on this job?"

"Absolutely."

"Fine!" I said. "I will make out the bill of sale and you can buy it at once."

"Buy it?"

"Certainly. Why not?"

"But I don't need to buy it. All I want to do is rent it for a couple of weeks."

"I am sorry," I said, "but my company sent me over here to sell tractors. I am allowed to give short demonstrations free of charge, but I am not permitted to do any renting. If you do not care to buy the machine I am afraid I shall have to take it away."

"This is terrible," said Luigi. "I can't let you take it away. But I hate to buy the thing outright when I will need it for only a little while. Surely we can reach some compromise."

At this moment we were interrupted. There was a knock at the door. Luigi opened it and admitted an important-looking gentleman, who handed him a paper, then bowed and departed. After Luigi had read the paper he scowled darkly and burst forth into a string of excited and indignant remarks in Italian. For several minutes he was too agitated to tell us what it was all about, but finally he calmed down and explained that the paper was some sort of court injunction which restrained him from moving away or disposing of any of the marble in his garden. The injunction had been obtained by Antonio Grimelli, who claimed that the marble was his property and was being illegally taken away by Luigi Bontade.

"This is an outrage," said Luigi. "It is a miscarriage of justice. It is unbelievable villainy. And how does Antonio get such quick action out of the courts? There must be something crooked about it."

"Maybe so," I said. "But it looks as if we were pretty well stuck. If we go ahead and move any more stones I am afraid that we will land in jail for contempt of court."

"But what can I do?" said Luigi. "That slimy toad will keep this matter in the courts all summer. He will keep all that refuse in my garden all summer. My business will be completely ruined."

"It looks pretty bad," I admitted.

"But I will have my revenge," snarled Luigi. "In the end I will get even with him. I am having my lawyer bring suit for one hundred thousand lire. He says I have a very good case. It will take time to try the matter, but in the end I am sure to win. And Antonio is not rich. When he pays me that money he will be ruined. My lawyer has already taken steps to tie up his property so that he cannot dispose of it. My revenge will come late, but it will be very sweet."

Luigi glared around the room and gritted his teeth ferociously.

"My goodness, Luigi," said Gadget, "you can certainly act bloodthirsty when you want to."

"All this trouble," said Luigi, "is the fault of Antonio. And I have a perfect right to be bloodthirsty—you will have to admit that."

"Do you really want to know what I think about this affair?" asked Gadget.

"Certainly," said Luigi.

"All right," said Gadget. "I will tell you. I think you are all acting like a bunch of idiots. The whole business makes me sick."

"What do you mean?"

"The way things stand at present," said Gadget, "everybody is going to lose except the lawyers. If you can't move that stone you, naturally, won't buy a tractor, and Mr. Botts and I will lose the sale. You won't be able to use the garden all summer, and you will lose practically all your business. And later, after you win your lawsuit, you will make a pauper out of Antonio."

"That is all true," said Luigi, "but what can we do about it?"

"I would suggest that you settle it out of court."

"That is impossible," said Luigi. "Antonio will not listen to reason."

"How much did you say you were suing for?"

"One hundred thousand lire."

"Would you be willing to take less if you could get it right away in cold cash?"

"Yes, I would."

"Would you be willing to take half as much—fifty thousand lire?"

Luigi considered. "Yes," he said at length, "I would take fifty thousand lire if I could get it right away and if Antonio would let me clean out the garden."

"Fine," said Gadget. "I am going over to see Antonio and try to make him listen to reason."

"It is impossible," said Luigi.

"I can try, anyway," said Gadget. "Goodbye."

She walked out the door, and we saw her cross the garden and enter the Café Grimelli.

This was a little before one o'clock. We did not see her again until a little after five o'clock, when she came gliding up to the front of the house in a gondola. This time she was accompanied by another Italian gentleman; he turned out to be Luigi's lawyer.

"I have been all over town," she said. "I have seen Antonio and Antonio's lawyer. I have conferred with my own lawyer and with Signor Sagredo. I have done a lot of business and had a lot of papers drawn up. And I have brought along your lawyer, Luigi, so that he can advise you."

We all followed Gadget into the dining room and stood around expectantly while she laid an imposing-looking legal document on the table.

"Here are the papers," said Gadget.

"What papers?" asked Luigi.

"This is an agreement," she explained, "between you, Luigi Bontade, and Antonio Grimelli. I have arranged everything. You get fifty thousand lire. Antonio agrees to permit you to clean the debris out of your garden at once. In return, you agree to waive all further claims for damages and to deliver the marble to Signor Sagredo, who has purchased it and who will send boats here to take it away. The agreement is easy to understand, and very much to your advantage. I would suggest that you and your lawyer go upstairs and talk it all over with your wife. Your lawyer tells me he will advise you to sign. When you are ready to do so, come back here."

Luigi picked up the papers and went upstairs with his lawyer.

"Holy Moses, Gadget!" I said. "How on earth did you manage to do all this?"

"It was rather simple," said Gadget. "When I saw Antonio I told him that I had come to help him out of his difficulties. I informed him that Luigi was bringing a damage suit against him for one hundred thousand lire, that Luigi had a very good case, and that he would undoubtedly win if it came to trial, but that I thought I could settle the matter out of court. I said that Luigi's chief desire was to get his garden cleaned out so he could continue his restaurant business. I told Antonio that if he would agree to lift the injunction which prevented Luigi from cleaning his garden, and if he would convey to me all his property rights in the ruins of the tower, I would agree to get from Luigi a release on all claims for damages.

"Antonio was very much surprised at getting such favorable terms. But he is a suspicious old bird, and refused to sign anything without the advice of his lawyer. Accordingly I agreed to meet him at his lawyer's office an hour later. In the meantime I visited the American Consulate and got the name of a reliable lawyer for myself. I called on him and had him draw up a legal contract to cover the verbal agreement which I had made with Antonio. Then the lawyer and I met Antonio at his lawyer's office, and Antonio signed up."

"Wait a minute," I said. "Did this agreement provide for Antonio to pay anything to Luigi?"

"No. He just conveyed to me whatever rights he might have in the wreckage."

"But I thought you told Luigi he was going to get fifty thousand lire?"

"I am coming to that part now. After I had signed the agreement with Antonio I took my lawyer and went over to see Signor Sagredo."

"Is he the man who was looking at the ruins this morning?"

"Yes," said Gadget. "He is an architect and builder—one of the best in all of Northern Italy. I had already consulted him this morning to get some information about the marble used in the tower."

"Why did you want any information about that?"

"It was like this," said Gadget. "After you went to the station to get the tractor, I got to looking at the ruins, and it seemed to me that there was a lot of very fine stone there. There were many pieces that were still unbroken. Some of them were carved and ornamented in a very beautiful way. I thought they might be worth a lot of money, and I wanted to find out."

"What good would that do you?"

"I wasn't quite sure. I thought that if the stones were valuable, Luigi might want to have them attached to insure payment of his claim for damages. I inquired at the American Consulate and they referred me to Signor Sagredo. He came over here and told me that the stone was all marble of the very finest quality. Apparently it had been brought all the way from the famous quarries of Carrara, which are clear over on the other side of Italy. He further stated that he would be glad to buy the stuff for his own use, and would pay eighty thousand lire for it. I told him I would take the matter up with the owners, and let him know about it later."

"So that is why you got Antonio to convey all that marble to you?"

"Exactly," said Gadget. "Antonio seems to be an incredibly dumb egg, and his lawyer is almost as bad. Neither of them had any idea how much the stuff was worth. After I left them I took my lawyer right over to Signor Sagredo's office, where we drew up and signed a contract by which I sold him my newly acquired rights in the stone for eighty thousand lire. I insisted on cash payment, and he gave it to me at once. I think he is getting this stuff pretty cheap, and for that reason was anxious to close the deal in a hurry. I had him give the money to me in two bank drafts, one for fifty thousand lire and one for thirty thousand lire. As soon as Luigi signs up I will hand him the one for fifty thousand."

"And what about the other thirty thousand?" I asked.

"I have that all figured out too," said Gadget. "Twenty-five thousand of it is the price of the tractor. The other five thousand will cover my lawyer's fee and leave a little extra to pay me for my services as diplomat and peacemaker. While I am over here there is a certain amount of shopping I want to do, so I will need a bit of pocket money."

"But I don't quite understand about the tractor."

"That is a little surprise for Luigi," said Gadget, "and here he comes now."

Luigi and Mrs. Luigi and the lawyer reentered the dining room.

"My lawyer and I," said Luigi, "are both amazed that you have been able to get all this money out of Antonio. We don't understand how you did it. But we will accept it with the greatest joy. Here is the agreement which I have just signed."

"Fine," said Gadget. "Here is a draft on the Banco di Napoli for fifty thousand lire. And I have some additional good news for you. Antonio—although he doesn't realize it himself—has been kind enough to buy for you the Earthworm tractor which we have been demonstrating. Marco, who is a splendid mechanic, will be able to run it for you so that you can clean out your garden in very short order. If you have no further use for the machine after you have finished this job, you can sell it, or keep it in the garden as a trellis for your rose vines, or use it any way you choose."

For a moment Luigi gazed at Gadget incredulously. Then, in his atrocious English, he made a few remarks which I will try to reproduce exactly. "Ah, *signora*," he said, "for longa time I have been told how in America da wife she bossa da husband. I have wonder why. Now I know. It is because American wife she have alla da brain in da family."

As I don't seem to think of any good answer to this crack, I will close.

Very sincerely,
ALEXANDER BOTTS.

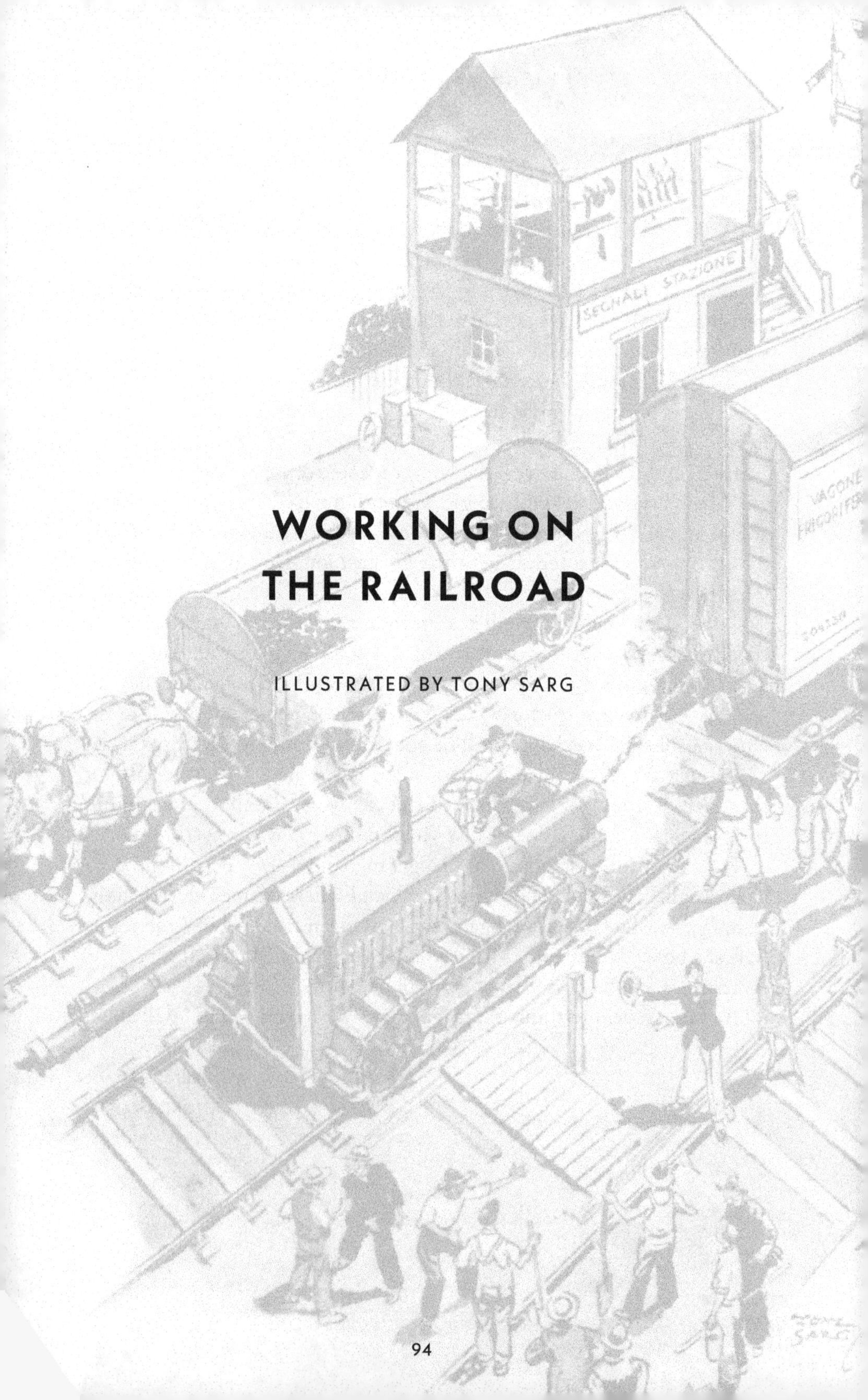

WORKING ON THE RAILROAD

ILLUSTRATED BY TONY SARG

EARTHWORM TRACTOR COMPANY
EARTHWORM CITY, ILLINOIS
OFFICE OF THE SALES MANAGER

APRIL 18, 1928.

MR. ALEXANDER BOTTS,
MARSEILLES, FRANCE.

DEAR BOTTS: We have just received your letter of April fourth describing the sale of an Earthworm tractor to a man in Venice. This letter, in connection with former ones covering your first month's work in Europe, indicates that you have completely failed to grasp the real significance of your mission.

We sent you over there to see if you could open up a European market important enough to justify us in establishing European branches—or possibly even a European factory.

We wanted you to get after people who might become large users of tractors, or to place machines in communities where there might be a chance for future sales.

And what have you been doing? You sold two machines to an American on the boat, and you sold several more to an American contractor who happened to be in France—all of which accomplished nothing in the way of introducing Earthworm tractors to the Europeans. Then you wasted many precious days constructing a makeshift cable railroad so that you could get a tractor in and out of an inaccessible hillside vineyard in the Alps. Any schoolboy would have realized that a cable car which would carry a tractor would also carry a mule; and yet you were surprised when the owner of the vineyard accepted the railroad and thanked you—an alleged tractor salesman—because you had shown him how he could use mules to much better advantage in his work than tractors.

All this was bad enough, but your latest report form Venice is the last straw. You go to the one city on earth which is totally unfitted for tractors. With the help of your wife and by the use of a certain amount of cheap cleverness, you unload one of your machines on a poor, simpleminded restaurant keeper who has no earthly use for it except a few day's work in cleaning some rubbish out of his little garden. And the worst of it is that you actually seem proud of this feat.

Of course, the selling of a tractor in Venice is a rather unusual stunt. It would also be an unusual stunt to sell a pair of snowshoes to a South Sea Islander. But the sale would be worthless from the point of view of building up a steady business. And anyone with any sense whatever should realize that the tractor you have placed in Venice is absolutely no good so far as future sales are concerned.

You should sell tractors to prominent farmers in prosperous agricultural regions where the neighbors will have an opportunity to observe them at work and be influenced to buy machines of their own. You should find prospects who really need tractors and who may become big users. Get after the people who are building and grading roads. Get after the people who have heavy hauling to do. See if you can't find some big lumber operators who ought to motorize their operations. But for heaven's sake lay off of these freak sales. And stay away from places like Venice.

Very truly yours,
GILBERT HENDERSON,
Sales Manager.

ALEXANDER BOTTS
EUROPEAN REPRESENTATIVE
FOR THE
EARTHWORM TRACTOR

GRAND HOTEL MIRAMARE & DE LA VILLA,
GENOA, ITALY.
WEDNESDAY MORNING, MAY 9, 1928.

MR. GILBERT HENDERSON,
EARTHWORM TRACTOR COMPANY,
EARTHWORM CITY, ILLINOIS.

DEAR HENDERSON: This will acknowledge your letter of April eighteenth, in which you see fit to make some rather ill-advised comments on my report of April fourth from Venice and in which you speak slightingly of my entire first month's work in Europe. I am very sorry indeed that you have assumed this attitude of petty faultfinding and narrow-minded querulousness. When I tell you all that I have been doing during my second month in Europe, you will see that I have been approaching the problem of making Europe tractor-minded in a very broad and comprehensive fashion. I have anticipated most of your suggestions about getting after the big markets. And, in consequence, your hasty criticism is entirely baseless. It is, in fact, most injurious, because your lack of confidence has discouraged and disheartened me at the very time when I need all my enthusiasm and self-assurance to carry things along.

I am now engaged in some very important sales negotiations, in the course of which I have met with some very annoying difficulties. If, in addition to these difficulties, I am to be burdened with the task of explaining and justifying myself to an unsympathetic and hostile sales manager at home, it will be more than even my courageous nature can stand.

I will not attempt to waste my time in answering your criticisms. I will merely request that from now on you do not cramp my style, and thus injure the Earthworm Tractor Company's future in Europe, by repeating them. Particularly, I would ask that in the future we regard this matter of the tractor in Venice as a closed incident. Having now stated my position, I will pass on to more important matters.

After leaving Venice, Gadget and I went directly to Milan. We had one of our tractors shipped up from Genoa, and we proceeded to show the farmers on the fertile plains of Lombardy what we could do. We got along very well, and I was continually impressed with my good fortune in having a wife as competent and efficient as Gadget. Owing to her knowledge of the Italian language, she was able to handle practically all of the talking, while I handled the tractor. We visited Pavia, Lodi, Novara and Monza. We even got up as far as Como. Gadget gave talks before various agricultural associations and groups of farmers, and we put on demonstrations of plowing, cultivating, stump pulling, and other things.

We found that Lombardy is particularly virgin territory for motorizing farming. Aside from the horses, mules, asses, oxen and hand labor, the only competition we ran into was a couple of cheap, out-of-date American tractors and a few incredibly clumsy machines made by a European firm. These pathetic contraptions are underpowered and expensive to run, are constantly getting stuck in soft or muddy ground, and are totally incapable of getting over the innumerable irrigation and drainage ditches with which the whole region is infested.

Most of the farmers we met seemed to be splendid fellows—intelligent, friendly, and very much interested in our beautiful Earthworm tractor.

They were all greatly surprised and delighted at its tremendous power and at the efficient way we drove along through mudholes, across ditches, and over seemingly impossible obstacles.

Unfortunately, however, labor and draft animals are fairly cheap over here. And the price of the tractor—on account of the freight and the heavy Italian duty—is pretty high. Consequently, the advantages of the tractor are not so apparent as in America, where labor is high and where machinery is cheap.

Everybody admired the Earthworm, but nobody wanted to be hurried into buying. So, after a dozen or more highly successful demonstrations, we decided to let the farmers of Lombardy think the matter over for a while. We will return later and get after them again. In the meantime we are in Genoa carrying on a campaign by which I hope to sell a large number of tractors to the *Ferrovie dello Stato*, or Italian railroad system, which, over here, is run by the government.

My idea is simple but magnificent. The railroad cars in Italy—as well as in the rest of Europe—are much smaller and lighter than our American cars. They are, in fact, so light that it is not necessary to use a big locomotive in switching them around; in the railroad yards and terminals all over the country horses and even gangs of men do this work. Obviously it should be done by Earthworm tractors. When this idea first hit me I made inquiries and found that at Genoa there are many piers with a tremendous network of railroad tracks running all over them. Consequently, there is a lot of switching of cars.

Exactly a week ago Gadget and I came down here, where, as you know, we have several tractors in storage. We got in touch with a bozo called Signor Galbi, who seems to be some sort of an official of the railroad. He told us he would be very glad to have us put on a demonstration. And for the past five days we have had a twenty-horsepower Earthworm busily engaged in pulling freight cars and switching them hither and thither all over the railroad yards and a half a dozen different piers. I have hired a young Italian to do the driving, and we have handled the work quickly, smoothly, and much more economically than it has ever been done before. Everything would be most encouraging except for some very curious difficulties which have suddenly arisen.

Yesterday afternoon Signor Galbi called Gadget and me into his office. "Your tractor," he said, "has been doing a very good job. It has been switching cars most efficiently." (Note: This was, of course, spoken in Italian. I am giving you an English translation so you will have a better chance to understand it.)

"We certainly are glad to hear you say that, *signor*," said Gadget, also speaking in Italian, "and now I suppose you will want to buy a dozen or more. We have the order blanks right here, and if you want we can sign you up right away."

"Not so fast," said Signor Galbi. "What I was starting to say is this: Your tractor has done very good work, but there have been so many criticisms of the machine and of your methods that I feel that I must ask you to stop."

"You mean you don't want to buy any tractors?"

"No," said Signor Galbi. "And I don't want you to demonstrate anymore."

"But what's the matter? Just what are these criticisms you talk about?"

"In the first place, you are causing trouble by the wages you are paying the man who is operating your tractor."

"His wages are perfectly fair," protested Gadget. "We are paying him one hundred lire a day. This is little more than five dollars, and is just what we would pay at home. Of course there are many tractor operators who get more than that, but they are experienced men. This bozo here is a good guy, but as far as Earthworm tractors are concerned, he is a beginner. One hundred lire a day is all he is going to get. And if anybody says we are trying to skin him, they don't know what they are talking about."

"You do not understand," said Signor Galbi. "It is not that you pay too little; it is that you pay too much."

"Too much?"

"Yes."

"What harm could that do?"

"It established a precedent," said Signor Galbi. "And it causes discontent among other workmen. Your mechanic has been boasting of his wages—which are just about four or five times bigger than they ought to be. Twenty or twenty-five lire a day would be plenty. If you go on paying these high wages, the workmen will get the idea that the job is worth that much. And if I bought any tractors and hired operators, I would have to pay the same exorbitant amount or else be accused of unfairness. And the thing would spread. Other workmen on other jobs would decide they ought to have more too."

"I see," said Gadget. "You are afraid everybody would strike for higher pay."

"No, they would not do that. Under the Fascist regime it is against the law for workmen to strike. But they would be dissatisfied. They would become inefficient. And instead of keeping their minds on their work, they would be thinking about their imaginary grievances. They would start appealing to the government for a higher wage scale, and they would make trouble everywhere."

"It seems to me," said Gadget, "that you are exaggerating this thing. Maybe we have paid our man too much. But you can't make me believe we have disorganized the labor situation of the whole part of Genoa."

"Fortunately," said Signor Galbi, "you have done very little harm as yet. But you have started something which may grow to alarming proportions if it is not stopped. There is nothing that demoralizes the working classes so fast as raising wages. You give them a little and they demand much more. They are never satisfied. Our only salvation is strict and severe discipline at all times. If we once let the insidious idea of higher pay get started, it will spread with great rapidity and soon poison the minds of the entire laboring class."

"Well, well," said Gadget. "This is most interesting. I wouldn't go so far as to say you have convinced me, *signor*. But we might as well play with you on your own terms. We will cut our operator's wages to twenty-five lire. If he doesn't like it, we will fire him and let you hire us somebody else at the reduced rate. And we will continue the demonstration until you are absolutely convinced that these tractors are what you need."

"No," said Signor Galbi, "I am afraid I shall have to ask you to stop—for the present at the least. There is another matter. Your tractors are too efficient: If we used them we wouldn't need so many men. Some of the people who now handle the switching of the cars would lose their jobs."

"That is something of a problem," admitted Gadget. "Naturally you wouldn't want to throw a lot of men out of work."

"They wouldn't exactly be thrown out of work," said Signor Galbi. "Under the Fascist regulations, employers are not allowed to lay off their employees. We would have to find them new jobs somewhere—that is the bad part of it."

"Why is that so bad? Aren't there plenty of other jobs on a big railroad system?"

"Yes, but it would make so much extra work for us in the executive part of the business. Shifting men and reorganizing the work is always troublesome. The men are satisfied with things as they are. I am satisfied. Why change? Why stir things up?" Signor Galbi leaned back in his easy chair and sighed.

"If that's the way you feel about it," snapped Gadget, "why did you let us start demonstrating in the first place?"

"That was a mistake on my part," he admitted. "You talked me into the tractor idea before I had thought much about it. But now that I have had time to consider the thing at my leisure I see that it is impractical. It would cause too much trouble. There would probably be a fight about the proper wages for tractor operators. And the quiet routine of our administrative office would be disrupted by the necessity of finding new jobs for a lot of men."

"But I am sure that all these difficulties can be overcome. If you would only let us continue our demonstration—"

"For the present," said Signor Galbi, "I think you had better stop. I will meditate further on this matter, and perhaps in the course of a week or two I may think of some way to handle things. In the meantime I will suggest that you keep your tractor in the storehouse of the Molo Vecchio."

"But, *signor*," said Gadget, "we have to waste a week or two sticking around here doing nothing. Why can't we thrash this thing out right now?"

"I do not care to argue the matter at present. I must have a lot of time for thought. If you cannot wait, the only thing we can do is to terminate the negotiations."

"In that case," said Gadget, "I suppose we will have to wait. Good afternoon! We will see you later."

We walked out of his office.

That was yesterday afternoon. I have been spending this morning at the hotel writing this letter, while Gadget has been down at the port looking the situation over and cautiously sounding out some of the men on their attitude toward tractors.

As I was writing the last few sentences, Gadget came in with some very interesting news which makes it necessary for us to leave at once for Milan and points north. I have no time to explain this new development, but before I end this letter I wish to make a few general observations.

You will note that I have given you a very complete report of our conversation with Signor Galbi. I did this, not because it was interesting but because it was so incredibly dumb that it ought to give you a very good idea of the difficulties we are meeting over here. Try to put yourself in my place and see if you can think of any way to make an impression on anything as solid as Signor Galbi's brain. If there was any sense to even a part of what he has to say, it might be possible to reason with him. But when he is so insensible to everything in the nature of a new idea, there isn't any place to start in on him. And yet, it is bozos like this that I must win over if I am to make any sales. I am in a very difficult situation.

And I wish you would remember this fact whenever you are tempted to make any hasty and ill-considered criticisms of what I am doing. Before you presume to bawl me out, please remember that I am one of the best salesmen you have ever had, that I know more about what I am doing here than you do, and that senseless faultfinding always does more harm than good. And whatever you do, don't let me hear anymore remarks about my activities in the city of Venice.

Very truly yours,
ALEXANDER BOTTS.

ALEXANDER BOTTS
EUROPEAN REPRESENTATIVE
FOR THE
EARTHWORM TRACTOR

ON BOARD THE GENOA-MILAN EXPRESS.
WEDNESDAY AFTERNOON,
MAY 9, 1928.

MR. GILBERT HENDERSON,
EARTHWORM TRACTOR COMPANY,
EARTHWORM CITY, ILLINOIS.

DEAR HENDERSON: We are on our way. We are about to pull off a coup d'état which will show these railroad people how good we are and may even convince a pessimist like yourself that we know our stuff. Just at the moment when things were most discouraging we discovered a wonderful opportunity. And we are taking advantage of it in our usual energetic way.

When Gadget went down to the port this morning she heard that there had been an accident on the main line of the railroad in the Alps just south of the Simplon tunnel. A dam had given way. The water had rushed down and washed out quite a section of the track. Apparently no one had been injured, but the railroad line was destroyed for some distance and all traffic stopped. The Simplon-Orient Express—one of the finest trains in Europe—had come through the tunnel, but had been halted at the washout. This was naturally considered very serious, and the railroad authorities were working frantically to repair the damage and to get traffic moving again.

As soon as Gadget heard this news she decided that our chance had come. When it comes to cleaning away rubbish and making quick repairs on a railroad, the Earthworm is right in its element. The regular wrecking cars work slowly and ponderously from one or both ends of the washout. They cannot touch the central part until they have particularly finished the repairs at the ends and laid new track so that they can get there. But the Earthworm can go right ahead over the toughest ground and do a lot of the preliminary work before the wrecking cars get anywhere near. If we could show the railroad how good we were in their work, it would give us a wonderful sales argument. Our machines would pay for themselves by their normal work in switching cars, and would be of tremendous advantage in times of emergency.

Gadget at once took the matter up with Signor Galbi, and he finally agreed to send me and Gadget and two tractors up to the scene of activities. I have a feeling that he fell in with our plan not because he thought it was any good—he is too dumb for that—but because he thought he could get a lot more rest if we were out of town. You have no idea how it annoys that man to have anybody around that is trying to bring to his attention anything remotely resembling a new idea.

We loaded the big sixty-horsepower machine and the thirty onto a sort of express car. These two machines had been in the warehouse here ever since they arrived from America two months ago. The express car, at Signor Galbi's orders, was coupled onto the afternoon northbound train. And now we are about halfway to the city of Milan. Gadget and I are riding right in the car with the tractors, so that there will be no danger of our becoming separated from them. We have with us written orders directing that upon our arrival in Milan we are to be transferred to the Paris Express. This, of course, will take us as far as the washout, where we will get off and proceed to astonish these Italians by doing our stuff. Gadget and I are both feeling fine. At last we are getting some action. And you can trust us to see that some real results will follow.

LATER, THURSDAY, MAY 10, 1 A.M.

All is well. We got into Milan after dark. After showing the written orders from Signor Galbi, we got our car attached to the Paris Express, which pulled out about midnight. We are now rolling merrily along through the night. And as we have a hard day ahead of us, we will now try to get a little sleep. At a time like this I certainly appreciate the fact that these Earthworm tractors have such soft and comfortable seat cushions.

LATER. 9 A.M.

We have run into some very hard luck. And it is all due to the incredible stupidity of Signor Galbi and the rest of these railroad officials.

When Gadget and I woke up this morning the train was just emerging from a tunnel. We peered out of the small window in our baggage car and were much pleased to observe that we were in the mountains. Far below us we could make out a little stream in a deep, dark valley, and far above us we had glimpses of tremendous snow fields and glaciers lighted up with a beautiful pink glow from the first rays of the rising sun. The train made a wide loop around the side of the little valley. It entered a town. There was a sign on the station that said Modane.

Gadget and I at once opened the door and got out. We strolled up and down the platform. We noticed that all the people from the passenger coaches were going into the station.

"I wonder how far we are from the big washout," I said. "We seem to be right up in the high mountains. We must be getting pretty close.

Gadget got out a map and began looking it over. "The name of this place seems to be Modane," she said. "But I can't find it on the map. We are supposed to enter the mountains at a place called Mergozzo. Then we pass Premosello and Domo d'Ossola. But I don't see anything called Modane."

"Maybe that is not the name of the town at all," I said. "Maybe it is just one of these signs like Dubonnet or Cinzano, or Sortie, or Herren or something like that."

"No," said Gadget. "It looks like the name of the town. And by the way," she continued, "what are those two bozos doing over there?"

She pointed toward our car and I noticed two men in uniform who were looking in at the tractors and carrying on some sort of discussion.

We walked over and asked them what seemed to be bothering them. One of them could speak English.

"Are those machines your property?" he asked.

"They are," I said.

"This is most irregular," he said.

"What seems to be the trouble?"

"We can't find any papers or any orders covering this shipment. There is no waybill, no invoice, no customs declaration. The conductor of the train here"—he pointed to the man beside him—"says this car was put on at Milan by a special order from one of the high railroad officials. But he says he has no information as to where it is going. It is most unusual."

"It is very simple," I said.

"Perhaps," suggested the man, "you have the papers covering this shipment?"

"I don't know what you mean by the papers," I said. "The only paper I had was the written order directing that this car be attached to the Paris Express, and I left that with the station master at Milan. But that might not make any difference. All I want to do is go along with this train as far as it goes. Is there any reason why I can't do that?"

"Plenty of reasons," said the man. "That order you spoke of may have been all right in Italy. But this train is going to Paris. The French railroad must have a waybill covering the freight charges, and I must have an invoice showing the value of this machinery, so it can be passed through the customs. Furthermore, you will have to have a passport in order to cross the frontier yourself."

"I have a passport," I said. "But that has nothing to do with this case. I am not going to France. I am stopping on the Italian side of the frontier."

"Oh, you are?" said the man. "Then why didn't you? Don't you know you are already in France?"

"Say," I said, "just where are we, anyway?"

"You are in Modane."

"Where is that?"

"Right here. You are in it now."

"No, no. I mean what country is it in?"

"In France, just across the frontier from Italy."

"You mean they have already fixed up the railroad line, and we have come right on through the Simplon tunnel?"

"You have just come through the Mont Cenis tunnel."

"I don't understand that at all," I said. "At Milan we got on the Paris train that was supposed to go by way of the Simplon tunnel."

"Exactly so," said the man, "that the Simplon line is closed on account of a flood, and the trains have been rerouted. Some of them went by way of the St. Gotthard tunnel. This train has been sent by the Mont Cenis route."

"And how far are we," I asked, "from the Simplon tunnel?"

"Several hundred kilometers."

"How many hours' ride?"

"I don't know," said the man. "Maybe eight or ten hours would get you there if you had good luck with connections."

For a minute or two Gadget and I were completely speechless.

"I am one of the French customs officers," continued the man in uniform. "If you will come with me into the custom's office, perhaps we can fix things up. You can make a declaration as to the value of this machinery, and pay the duty, and I think we can let you go on in spite of the fact that you do not have proper papers."

"I just told you," I answered, "that I don't want to take these things into France. I want to go right back into Italy and get up to the Simplon tunnel, where they had the big flood. These two tractors are to be issued in repairing the line."

"Oh!" said the man. "You came this way by mistake?"

"Yes."

"And you want to go back to Italy again?"

"That's what I have been trying to tell you."

"Come with me," he said.

He took Gadget and me into the customs office and explained everything to his chief. The chief called in several advisers and we had a long and intricate discussion in French, English and Italian. At first the chief was going to charge us duty on the tractor because the frontier is in the middle of the Mont Cenis tunnel and consequently the tractor had already been brought into French territory. But after Gadget had argued with him for ten or fifteen minutes he consented to let us off, provided we took our machines straight back to Italy. After this it took fifteen or twenty minutes to persuade the train conductor to uncouple the car. Then we waited an hour or so until a train bound for Italy came along, and after another interminable argument we got our car hitched on and started back toward the frontier again.

I have written this report partly while waiting at the station in Modane, and partly since we got underway. We have now just completed the long eight-mile trip through the big tunnel, and we are on Italian soil once more. We are due back in Milan at a little after three this afternoon. If we don't run into more trouble there, we ought to be able to catch a train which would get us to the washout sometime this evening. We may not be able to do much tonight, but we should be on the job first thing in the morning.

LATER THE SAME DAY. 1 P.M.

Most stupidity, more trouble, more grief. When our train stopped at Bardonecchia, which is the first town on the Italian side of the Mont Cenis tunnel, a couple of bozos in uniform came up to our car, looked at us, looked at the tractors, and then started a long harangue in Italian. I am getting so I don't like these people in uniform. As Gadget understands the language, I let her deal with them. She has given me a complete account of everything that was said, and I will let you have the main points.

It seems that the men were Italian customs officers, or members of some sort of a border patrol. First they looked over our passports. Then they wanted the papers covering the shipment of the tractors. Of course we had no papers. And neither had the conductor of the train.

"What?" said one of the men. "You have nothing to prove that you have paid duty on these machines?"

"We are not paying duty," said Gadget.

"What? You are bringing these machines into Italy without declaring them—without paying the legal duty?"

"We are not bringing them into Italy."

"But they just came through the tunnel, didn't they?"

"Yes."

"Then they came from France."

"No, they just came from Modane."

"Modane is in France."

"I know," said Gadget, "but it is so near the frontier that it ought not to count."

She then gave them a long explanation of exactly what had happened. She argued that as the tractors had never been officially admitted to France, they had never been officially out of Italy. As Gadget had talked up this very well, and as she is a wonderful persuader, we might have got by with it except for another legal point that bobbed up. As soon as the officials heard that we had gone through the tunnel last night, they made a careful examination of our passports.

"If what you tell me is true," said the one who was doing the talking, "you have committed a very serious offense. The law provides that no one may cross the frontier to leave Italy without the official *uscita* of the government. Our men inspect each train before it goes through the tunnel. They examine the passports of the passengers and stamp them with the official media. But your passports are not stamped. Why?"

"We never saw your old inspectors," said Gadget. "We were riding in the baggage car."

"I see," said the man. "You were in hiding. It is a very serious offense and I am compelled to put you both under arrest. We will take your car off the train and hold it here until we have official instructions as to the disposition of your property."

They proceeded at once to leave our car backed onto a siding and uncoupled. The rest of the train went on its way, and the inspectors took us to their headquarters, where they apparently intended to lock us up, or chain us to the wall or something.

But Gadget, as I remarked before, is a wonderful talker and has a very smooth way of handling people. And, fortunately these two men in uniform were pretty decent chaps, in spite of their cockeyed ideas about what the law required. Gadget succeeded, after much lengthy conversation, in convincing them that we were innocent of any criminal intentions, and were merely the victims of an unfortunate mistake. They finally agreed to forget about our passport irregularities, and they said that if we would go back to Modane, see the Italian customs officers who are stationed there, and get them to give us a written statement certifying that our tractors had really come from Italy, and had never been officially admitted to France, they would drop all action against us and let us take possession of our property once more.

As this was the best arrangement we could make, we had to agree. We are now waiting for the afternoon train to Modane, and we shall soon be on our way back through the tunnel again. To obviate any possible difficulties with the railroad, we have sent a long telegram to Signor Galbi in Genoa, explaining the situation and asking him to send the necessary orders to the railroad officials so that we can proceed to Milan and then to the Simplon without delay.

Modane.
Later the Same Day. 7 p.m.

More delay. We left Bardonecchia late in the afternoon. This time we got the official *uscita* stamped on our passports. We reached Modane about suppertime, and found that the Italian customs man we have to see had gone off duty. We can't get anything done until tomorrow, so we have taken a room at a little hotel here. This thing is beginning to get on my nerves. How I hate these birds in uniform. When I am just raring to go, with all this important work waiting for me up by the Simplon tunnel, it is naturally most irksome to put in days and days riding back and forth through this Mont Cenis tunnel. If we don't get some action tomorrow, there will probably be dispatches in the papers stating that an American traveler by name of Botts suddenly went insane in the Alps near the Franco-Italian frontier and shot a number of Italian officials.

May 11, 9 a.m.

It is now the next morning. The Italian customs man gave us the statement we wanted, and we are now on the train riding through the tunnel, once more on our way to Bardonecchia.

The guidebook says the construction of this tunnel was begun in 1857, during the reign of the Emperor Louis Napoleon of France. It was finished in 1871. It is almost eight miles long, is big enough for two tracks and is considered a great engineering achievement. But it doesn't thrill me at all. After we get out of it, the whole thing can cave in for all I care. And I don't want to see or hear anything more about it for the rest of my life.

Bardonecchia.
Later. 10 a.m.

I don't know why I keep writing you this series of reports. Gadget and I seem to be engaged in a lot of activities over here, but all to no purpose. Our proceedings have been so futile and so completely pointless that I can't imagine anyone taking any interest whatsoever in reading them. However, as you people are paying the salary and expenses, perhaps you may have a mild desire to know what we are doing, or not doing.

We arrived at Bardonecchia about half an hour ago. We found a long telegram, collect, from Signor Galbi. Being translated it reads about as follows: "Am sending orders for car with tractors to return to Genoa. Simplon line now repaired. We are advised that much of the work was done by a salesman demonstrating some other tractor, who has so impressed officials there that they will buy his machine, not yours."

So that is that. We are starting back to Genoa by the next train. I have not yet shot any of these birds in uniform, but I don't know how long I can control myself. I will write you from Genoa and let you know my plans for the future.

Yours sincerely,
Alexander Botts.

ALEXANDER BOTTS
EUROPEAN REPRESENTATIVE
FOR THE
EARTHWORM TRACTOR

GRAND HOTEL MIRAMARE & DE LA VILLE,
GENOA, ITALY.
MAY 12, 1928.

MR. GILBERT HENDERSON,
EARTHWORM TRACTOR COMPANY,
EARTHWORM CITY, ILLINOIS.

DEAR HENDERSON: We got back to Genoa last night. This morning, in a rather despondent state of mind, we called on Signor Galbi.

"I am glad to see you," he said. "There are a couple of gentlemen here who wish to talk to you."

He took us into an inner office and introduced us to a very large, very pompous and very important-looking man. He was not very tall, but he was heavily built, and had a brown mustache and a luxurious brown beard, which was evidently very well taken care of. It was neatly trimmed and of a beautiful silky texture which indicated much brushing and combing.

The man's name was General Brera. He came from Milan, and was one of the big brass hats of the railroad company. Beside him stood a tall and muscular young man who looked vaguely familiar. It was not until he shook hands with me that I recognized him.

"Well, well, well," I said, "if it isn't good old Marco himself—the demon tractor driver."

Note: It was indeed Marco. He is the young Italian whom I taught to drive the tractor which I sold a month or so ago to Signor Luigi Bontade in Venice. He is a very bright lad and he talks English fairly well.

"I have come," said Marco, "to ask you for a job as salesman. And I have brought you a prospect."

"Tell me all about it," I said.

"I will," said Marco. "About a week ago Luigi finished clearing up his garden and had no more use for the tractor. I had saved up a little money, so I bought it very cheap and took it to Milan to see if I could sell it to a cousin of mine who owns a farm near there.

"He introduced us to a very large, pompous man."

"At Milan I heard that the railroad needed to make some quick repairs on their line near the Simplon. I got them interested in my tractor. I shipped it up there and it did such good work that they have bought it. General Brera was so impressed that he wishes to buy a half dozen. If these work out well, he may get a lot more."

"It seems incredible," I said.

"Not at all," said Marco. "Your Earthworm is such a good machine that I think I can sell quite a number of them in other places. I want you to give me a job as salesman."

"Marco," I said, "this is more than I seem able to grasp all at once. Suppose we sit down and go over everything slowly and carefully, so that what is left of my mind can have a chance to get hold of it."

We sat down. The general told us clearly and simply what he wanted, and it was a delight to talk to him. He is a man of action.

He was confident that he could easily handle any labor or wage difficulties that might result from using tractors. And his thinking is as clear and sound as Signor Galbi's is muddy and devious. Perhaps that is why he is one of the main guys in the railroad, while old Galbi has merely a second-rate local job. Within a few minutes the general had signed up an order for six machines.

Then I hired Marco as our Italian salesman. I know I have no authority to do this, but I have gone ahead anyway. The company needs this guy.

His wages will be three thousand lire a month, and I want you to cable me that amount at once, so that he can get his first month's pay promptly on June first.

In conclusion, I wish to take back the request I made in a former letter that you refrain from making any further mention of the tractor which I sold in Venice. Hereafter you may talk about it as much as you wish.

And in this connection I would like to refer you to the famous remark made many years ago by a certain gentleman of New York City and Charlottesville, Virginia. You will remember that this gentleman, under circumstances slightly similar to those in which I find myself, telegraphed his family the simple rhetorical question, "Who's loony now?"

I beg to remain your very respectful servant,
ALEXANDER BOTTS.

THE NEW MODEL

ILLUSTRATED BY TONY SARG

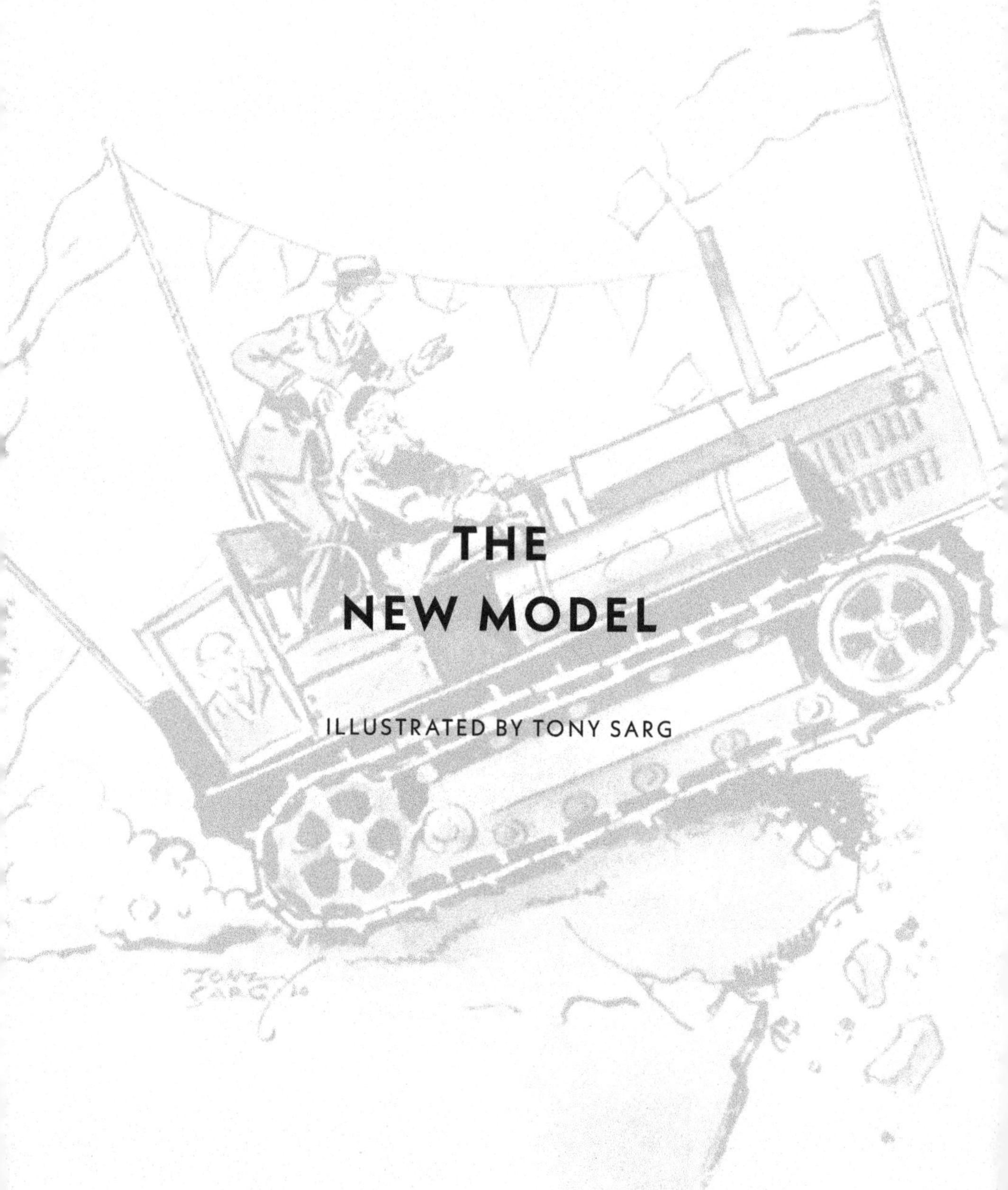

ALEXANDER BOTTS
EUROPEAN REPRESENTATIVE
FOR THE
EARTHWORM TRACTOR

GRAND HOTEL MIRAMARE & DE LA VILLE,
GENOA, ITALY.
WEDNESDAY, MAY 16, 1928.

MR. GILBERT HENDERSON,
EARTHWORM TRACTOR COMPANY,
EARTHWORM CITY, ILLINOIS.

DEAR HENDERSON: I have great news for you. I believe that I am about to open up a new and tremendously important field for the sale of Earthworm tractors. Since my arrival in Europe more than two months ago, I have, as you know, done pretty good work. I have sold an even dozen machines—a truly remarkable number, considering the difficulties which have confronted me. But all my previous achievements are about to pale into insignificance as compared to the tremendous proposition I am now undertaking.

It was last Monday morning that a gentleman called at the hotel, sent up a card bearing the name of Vladimir Krimsky, and asked to see me on important business. Gadget and I at once descended to the lobby, where we found a tall and very impressive gentleman awaiting us. He was about forty years old, smooth-shaven, and dressed in a neat and rather expensive-looking business suit.

"Pleased to meet you, Mr. Krimsky," I said. "I am Mr. Alexander Botts himself." I then presented Gadget. "This is Mrs. Botts," I said. "She speaks French, German and Italian. All you have to do is pick your language, and she will translate it so I can understand it."

"It is an honor to meet you, Mrs. Botts," he replied, with a low bow. "And you, too, Mr. Botts. I fancy we shall get along very well speaking English. I know it reasonably well."

"You sure do," I said. "You speak it swell."

"Thank you," he said.

"And what can I do for you today?" I asked.

"I wish to see you on a matter of business. I understand that you are the European representative for the Earthworm tractor."

"That is correct," I said.

"I am the representative," he said, "of the Ukrainian Cooperative Society, which handles most of the buying and selling for the farmers in Ukrainia."

"Where is that?" I asked.

"It is in Southern Russia. I have been sent here to Italy by the Soviet Government to handle any business which my cooperative society may have in this region."

"Well, well," I said, "are you one of these Bolsheviks that I have been hearing about?"

"I am a member in good standing of the Communist Party," he said, smiling pleasantly.

Note: I will have to admit that this surprised me a good deal. I had always supposed that Bolsheviks and communists were strange, uncouth creatures with bushy whiskers, who went about throwing bombs in all directions. This bozo was, as I have said, very well dressed; he had no whiskers at all, he seemed perfectly harmless, and his manner was polished and suave. It was, therefore, something of a jolt to hear him calmly tell me he was a communist. However, as I am naturally polite, I did not kid him about it.

"You say you are interested in tractors?" I asked.

"I am," he said. "I was in Venice some weeks ago and happened to observe a machine at work clearing up the ruins of a large tower which had recently fallen down. This machine was doing the work so much more easily and quickly than any other tractor I had ever seen that I interviewed the owner."

"Whose name," I interrupted, "was Luigi Bontade. Yes, Mr. Krimsky, I sold that tractor to Luigi myself—with a little help from my wife here. What did he have to say about it?"

"He said that I was correct in thinking it was a very superior tractor. He told me that it was called the Earthworm, and that it came from America. He gave me your name and address, and said that you were the European representative of the company that makes it."

Note: As you see, all this new business is sprouting directly from that sale in Venice. I don't want to rub it in too much, but I cannot help delicately reminding you that not so long ago you wrote me a letter bawling me out for spending so much time in Venice, "because," as you said, "it is obviously such a poor market for tractors that any sale in Venice can never help build up future business." Well, all I can say is that my native sense of courtesy prevents me telling you what I think of your opinion.

I assured Mr. Krimsky that I was indeed the European representative of the great Earthworm Tractor Company. "If you want to buy any of our machines," I said, "I am the guy that will sell them to you."

"Very good," he replied. "I have authority from my superiors to offer you a most favorable proposition. I want you to come with me to Ukrainia with one or two tractors and demonstrate them in our farming villages. If they prove satisfactory, they will be bought by the local branches of the Ukrainian Cooperative Society, and orders will probably be placed for several more. If they are not satisfactory, you will lose nothing, as I am authorized to pay all your expenses, coming and going."

"That sounds fair enough," I said. "Do you really think there is any market for tractors in Russia?"

"I know there is. Russia is tremendous country, with a tremendous population. There are enormous tracts of fabulously rich farm lands which are capable of raising more grain than all the rest of the world put together. All we need is modern methods and modern machinery. If your tractors prove to be as good as I think they are, it is possible that before the end of the year our society may buy several hundred of them."

"Pardon me," I said. "Did you actually say several hundred?"

"I did," said Mr. Krimsky. "But that is only a beginning. These few hundred machines would be merely for some of the Ukrainian peasants on individual farms who wish to make group purchases of farm machinery through our society. But when the collectives get going we will begin to do some real heavy business."

"Collectives?"

"Yes," said Mr. Krimsky. "Within a year or two the Soviet Government—which is the most progressive government on earth—will have all the land organized into large collective farms operating on a quantity-production basis. These collectives will all be controlled by a single administrative head, and will form the most stupendous, most efficient and most beneficent agricultural system ever known to man. When this scheme gets started we shall need thousands of tractors—yes, tens of thousands."

"You don't know how you interest me," I said. "You seem to have a reasonably large proposition here, and as long as you want to let me in on the ground floor, I will accept your offer. Business in Italy is at the moment a bit dull, so there is no reason why my wife and I can't both go."

"I am delighted to hear you say so."

"We will take along a new and much improved thirty-horsepower Earthworm which, by great good luck, has just arrived from America.

We will also take one of the old-model sixty-horsepower machines. And we will take a plow. When do we start?'"

Mr. Krimsky suggested that we embark on the *Santa Lucia*, an Italian freight boat, which leaves next Saturday for Odessa. Gadget and I said we were willing, and it was so arranged. Mr. Krimsky gave us credit references and explained to us various details, such as passport and visa requirements, and so on. Then he bade us a very courteous good morning and told us he would see us on the boat.

We are leaving Marco Manzione, the young Italian whom we recently hired as a salesman, to look after any business that may come up here in Italy. And on Saturday we are sailing for the fabled land of the Muscovites in high hopes of accomplishing new and dazzling feats of large-scale selling.

As ever, your vigorous and efficient salesman,
ALEXANDER BOTTS.

ALEXANDER BOTTS
EUROPEAN REPRESENTATIVE
FOR THE
EARTHWORM TRACTOR

ON BOARD STEAMSHIP *SANTA LUCIA*,
SUNDAY, JUNE 3, 1928.
FIFTEEN DAYS OUT FROM GENOA.

MR. GILBERT HENDERSON,
EARTHWORM TRACTOR COMPANY,
EARTHWORM CITY, ILLINOIS.

DEAR HENDERSON: We are due in Odessa tomorrow morning. It has been a long, slow voyage, but I can assure you that Gadget and I have not been wasting our time. We have, in fact, been improving each and every shining hour by feeding Mr. Krimsky large doses of sales talk on the subject of the Earthworm tractor, laying particular stress on the advantages of the new thirty-horsepower model. Mr. Krimsky—although a highly intelligent man—is a red-hot Bolshevik, and whenever we start talking tractors, he tries to twist the conversation to the subject of communism, so that he can give us an inspirational sermon on what a good thing it is.

Why he should want to talk about mere political affairs when he can listen to us discussing tractors, I don't know. But that seems to be the way he is.

Only this morning I had occasion to refer to the fact that our new thirty-horsepower model is a real masterpiece of engineering art. "Every useless part is eliminated," I said. "Every needed part is so placed and so coordinated with every other part that it can work to greatest advantage. As a result of all this, the new model is probably the world's finest example of harmony of design."

"An even finer example," interrupted Mr. Krimsky, "is the communist state. We have eliminated all useless and parasitic elements, such as the predatory bourgeois classes who wrongfully exploit the labor of others. And we are organizing the industrial and agricultural life of the country into a single rationalized system, which is destined soon to lead the world in efficient production."

"Did we tell you about our new gasoline tank?" asked Gadget. "It is set in such a position on the machine that we get rid of all troublesome elements, such as—"

"Speaking of troublesome elements," said Mr. Krimsky, "Lenin has well said that religion is the opium of the people. Now, in the communist state—"

"However," continued Gadget, "we have not stopped with eliminating useless parts. In our new-model Earthworm we have greatly improved the remaining necessary parts. Wait till you see our latest-type precision-movement, super-delicate steering mechanism. One touch from a baby's hand, and the entire machine—weighing more than five tons—turns around in a space no larger than a silver ten-kopeck piece."

"Another good point," I continued, "is the new system of velvet-grip brakes." And I then proceeded to give him a highly interesting and instructive half-hour lecture on the advantages of this feature of the new model.

From the above you can get an idea of what has been going on throughout this entire voyage. Mr. Krimsky is constantly attempting to discuss his favorite topic of communism. But Gadget and I, working as a team, have always been able to overwhelm him with an irresistible flood of tractor sales talk. We have got him so saturated with our ideas that after we have made our demonstration he will just automatically start urging his associates to buy as many tractors as they possibly can.

I have high hopes for the future.

Yours,
ALEXANDER BOTTS.

ALEXANDER BOTTS
EUROPEAN REPRESENTATIVE
FOR THE
EARTHWORM TRACTOR

HOTEL LONDONSKAYA,
ODESSA, U.S.S.R.
MONDAY, JUNE 4, 1928.

MR. GILBERT HENDERSON,
EARTHWORM TRACTOR COMPANY,
EARTHWORM CITY, ILLINOIS.

DEAR HENDERSON: Early this morning we landed at Odessa, which is a beautiful town on a high bluff, overlooking the sea. It has wide, well-paved streets, handsome buildings and a very fine harbor, with stone and concrete piers. There are many grain elevators and big oil tanks. The railroad tracks here run right out onto the piers, just as they do at Genoa and other European ports. Mr. Krimsky explained that this makes the handling of freight very easy, and is a great improvement over the medieval methods still used on many of the piers at New York.

We got the two tractors unloaded very promptly, and we have put in a busy and exciting day. Mr. Krimsky had, of course, notified his friends that we were coming, and it had been arranged that the new-model thirty-horsepower tractor was to be shipped this evening to an outlying village called Usk, where we are to put on a demonstration.

As this plan left us the day unoccupied, and as I am always eager to advertise the Earthworm tractor, I decided to inaugurate a combination parade and demonstration for the benefit of the citizens of Odessa. Unfortunately, this demonstration was not a complete success, but I will give you a brief account of it, so you can see that I did everything all right, and that the trouble we ran into was in no way my fault. In fact, my skillful driving was the only thing that prevented a holocaust.

As usual, I made my plans with the greatest skill and forethought. And I executed them with all my usual energy and efficiency. As soon as we had attended to the formalities at the port and had our passports and papers properly fixed up, Gadget, with the baggage, took an automobile to the hotel. Mr. Krimsky and I grabbed a venerable horse-drawn cab and speeded up to a hardware store on the Uliza-Lenina, where we bought a large can of high-gloss, quick-drying, red enamel. Then we shopped around and got a number of banners and pictures, threw them all into the cab, and raced back to the port. Here, for a few kopecks, I was able to hire a number of dock laborers to repaint the new model. By the end of the morning the work was completed, and the tractor was indeed a remarkable thing to look at. Never have I seen such a swell paint job. The entire machine fairly glowed a rich, warm crimson. This color, of course, was intended as a delicate compliment to the Soviet Government.

As soon as the paint was dry, I fastened on the banners and other paraphernalia. On top of the radiator I had a large red flag. At the rear I had a beautiful banner with a lot of Russian letters on it meaning, "Long Live the Revolution." On the side of the machine I put a number of smaller red flags and banners of various kinds. And as the crowning artistic touch I affixed two large colored pictures of Lenin and Karl Marx, who seem to be the chief patron saints of the country. As Karl Marx had by far the best-looking whiskers, I gave him the place of honor up in front. By the time I had finished the decorating, Gadget had returned from the hotel. She was, naturally, delighted with what I had done, and we at once asked Mr. Krimsky to take a ride with us.

"We are going to drive all over town," I said, "and let these bozos see what a real tractor looks like."

At first he was a little doubtful as to whether it would be legal for us to have a parade without securing a permit from the police, but I finally persuaded him that one tractor wasn't a parade and it ought to be just as lawful as three people riding in an automobile. He finally agreed; we all three climbed up into the broad and handsomely cushioned seat and I gave her the gas. With a splendid roaring and clanking I drove over the granite pavement of the pier, and then continued along the waterfront and past the coal port, looking for a street to take me up the hill into the town.

Almost at once, however, I discovered something far better than a street. Directly in front of me I was delighted to observe a most magnificent flight of steps. These steps are broad and beautiful. They are made of stone; there must be several hundred of them; and they lead straight up the bluff to the town.

"This is a historic spot," said Mr. Krimsky. "During the evil reign of the Czar a company of soldiers fired from those steps into a crowd of rebellious citizens. It was a terrible massacre."

"That is most interesting," I replied, "because the place is about to be the scene of another historic event—the first Russian demonstration of the climbing abilities of the great Earthworm tractor."

"What are you going to do?"

"I am going to give the people of Odessa a show," I said, "and I'm going to give them a good one."

I drove to the bottom of the steps. I shifted into low. I started up.

"Stop!" yelled Mr. Krimsky. "You can't do this!"

"Sure I can!" I yelled back. And I kept on.

It was a rough ride, but the tractor rolled along up powerfully and surely, and I would have made it easily if it had not been for two very unfortunate occurrences. The first of these was the sudden appearance of a half a dozen gentlemen in uniform. Apparently they were policemen. They stood at the top of the steps, directly in my path, and held up their hands as a signal for me to stop. I couldn't go on without running over them. So I threw out the clutch, pulled on the brake and shifted the gears into neutral.

And then came the second unfortunate occurrence. The newfangled velvet-grip brakes refused to hold, and we started rolling backward. Mr. Krimsky let out an ear-shattering screech and leaped from the machine. Gadget let out an even louder yell, but stuck by me. And I would have made more noise than both of them, except that I was too busy. I pulled and heaved at the gear-shift lever, but the gears were already spinning so fast that I couldn't get them meshed. Faster and faster we rolled. The emergency brake seemed to be doing no good at all, but, by bearing down with all my strength on the two foot brakes, I found I could slow up our progress a little and do a certain amount of steering. Looking back over my shoulder, I could see the port, apparently very far away, and at a dizzy distance beneath me. A large crowd, which had gathered very rapidly at the bottom of the steps, was now dispersing even more rapidly. And it was just as well, or the famous steps of Odessa might have been the scene of another massacre. As it was, I almost hit a dozen or more fleeing pedestrians and came within inches of demolishing three dogs and a baby carriage. Down, down we went, for what

"I pulled and heaved at the gear-shift lever, but the gears were already spinning so fast that I couldn't get them meshed. Faster and faster we rolled."

seemed like hours, although I don't suppose it was really more than thirty seconds. At last we reached the bottom, rolled out over the flat pavement for a short distance, and came to a stop.

Fortunately, nobody was hurt, and, fortunately, the tractor was all right. The motor was still running, so I started back for the pier. But I was soon blocked by the crowd which had gathered. I stopped. Mr. Krimsky came running down the steps and climbed aboard just as another lot of policemen accosted us. They seemed very angry indeed, but Mr. Krimsky talked to them most effectively and showed them his official papers. They finally stepped back and shooed the crowd away from in front of us, and I was able to continue to the pier.

Here Mr. Krimsky bawled me out good and plenty for attempting to go up the steps. And for once in my life I was pretty near speechless. I couldn't say anything except to admit that the tractor was at fault.

"These improved brakes," I said, "are the finest in the world. But they are so new that apparently the boys at the factory don't understand them yet. They left them too loose. But I will readjust them right away, and we will never have an accident like this again."

"I am sure of one thing," said Mr. Krimsky, still somewhat sore, "you won't have any more accidents on those steps. If I ever see you driving in that direction again, I'll knock you over the head with one of your own big track wrenches."

"And I wouldn't blame you," I said.

At once I got busy and tightened up the brakes, after which I loaded the tractor and the plow onto a freight car. They will go out tonight, and Gadget, Mr. Krimsky and I will follow on the passenger train tomorrow morning. And by tomorrow afternoon we hope to be starting our great demonstration in the little village of Usk, which is on the banks of the River Bug, about a hundred and fifty *versts*—a hundred miles—northeast of here.

In conclusion, I wish to ask you to present my compliments to the final inspection department at the plant, and tell them that they are a disgrace to a respectable company. If they let any more machines come through without adjusting the brakes, I will go out there when I get home and fix them so their necks will need adjusting.

Yours,
ALEXANDER BOTTS.

Alexander Botts
European Representative
for the
Earthworm Tractor

Usk, Ukrainia, U.S.S.R.
June 5, 1928, 7 p.m.

Mr. Gilbert Henderson,
Earthworm Tractor Company,
Earthworm City, Illinois.

DEAR HENDERSON: Here we are, and everything is going fine. We have had a few minor accidents and mishaps, but nothing serious enough to interfere with our great selling campaign.

Gadget and I and Mr. Krimsky took the train out of Odessa this morning. The railroads here seem very good, in spite of the fact that the rails are clumsily spaced considerably farther apart than in America. We reached Usk about noon and found that the tractor had already arrived, or rather part of it had. Somewhere on the way somebody had stolen the carburetor and the magneto. Naturally, this made me very sore, particularly as Mr. Krimsky had been lecturing us most of the morning on the splendid loyalty of all the people to the Soviet Government. According to him, all the working class was doing everything possible to cooperate with the government in the efforts to improve the industry and agriculture of the country.

"It seems to me," I said, "that there is at least one guy in this country that has an awful funny idea of cooperation. He certainly has helped us a lot by swiping our carburetor and magneto."

"This is indeed most regrettable," said Mr. Krimsky, "but, of course, the crime could not have been committed by one of the working class. It was probably done by a former member of the so-called upper or middle classes as an expression of his counterrevolutionary bourgeois ideology."

"Oh, I see," I said. "That makes it all right, I suppose."

At this moment a couple of railroad officials arrived, leading a huge lout of a peasant who had a face like a chimpanzee. I was interested to observe that this uncouth creature was carrying in his hairy paws the missing magneto and carburetor. At once there was a long argument in Russian, or Ukrainian, or something—which, of course, neither Gadget nor I understood—and then Mr. Krimsky explained the matter to us.

It appeared that the strange-looking missing link who had taken our stuff was a good communist by the name of Ivan Tschukovsky. His act was due not to any counterrevolutionary bourgeois ideology but rather to a mistaken attempt to assist the Soviet Government in making a success of National Junk Week.

"And what," I asked, "is National Junk Week?"

"The government," explained Mr. Krimsky, "in order to promote the well-being of the nation, has designated certain weeks which are to be devoted to concentrated efforts on behalf of various worthy activities. We have had National Defense Week, National Health Week, and many others."

"Yes, yes," I said. "We have them in America. We have Be Kind to Animals Week, Eat More Cheese Week, and I don't know what all. It is a wonderful idea."

"It is, indeed," said Mr. Krimsky. "And it happens that this is National Junk Week, which is to be devoted to the prevention of waste all over the Soviet Union. All good citizens are urged to gather up and bring in to designated depots all junk which would otherwise go to waste, such as scrap iron, old brass, rags, bottles and so on."

"And Comrade Tschukovsky," I said, "like the good communist he is, was trying to perform his daily good deed by converting my perfectly good tractor into junk?"

"Apparently that was it," said Mr. Krimsky. "He made a mistake. But he meant well, so everything is all right."

"Yes," I said, "but I hope there won't be too many of these errors."

As Comrade Tschukovsky shuffled away, I put the stuff back on the tractor. I unloaded the machine and the plow, and then we all climbed into the seat and drove down the main street of the village. Gadget and I were both much surprised at how primitive the place was. The village consisted of perhaps a hundred little shacks built of logs—something like the old American frontier log cabins. The roofs were straw. The main street was unpaved, and dry and dusty under the hot sun. It was very different from a French or Italian village; the only point of similarity being the manure piles and the pigs and chickens.

The entire population turned out to watch us as we went by. They appeared strong and healthy and very friendly, but a bit poverty-stricken.

"I thought you told me," I remarked to Mr. Krimsky, "that your new communistic system was going to make everybody rich. I don't see much evidence of it here."

"Give us time," he replied. "In another ten years you won't know this place."

I drove on. Near the end of the street Mr. Krimsky made some inquiries, and we finally stopped in front of the house of a man called Chipkoff, who was to be our host and who was to be taught how to drive the tractor.

I could see right away that Comrade Chipkoff was a regular guy, and that he and I were going to get along. Of course he couldn't speak English, but he had an intelligent face and I liked his looks.

We followed him into the home. It had three rooms—instead of just one, as in most of the other houses of the village—and it was all scrubbed up clean, and very neat. There was a new wooden floor in the main room, and a big stove made out of bricks. There were good-looking copper cooking utensils hanging around, and in one corner I saw an American sewing machine. Comrade Chipkoff introduced us to his wife and two children. They were a fine-looking family.

Mrs. Chipkoff set out a meal for us—simple, and plain, but very good—and afterward I gave the good comrade a driving lesson. Mr. Krimsky went along as interpreter, and we made splendid progress. My pupil had never before seen a tractor, but he had such a quick mind that he learned with great rapidity, and I am going to let him drive in the big plowing demonstration which I plan to put on tomorrow morning.

If one of their own people drives the tractor, I think it will make a better impression upon the inhabitants of the village. And it is the inhabitants—practically all of whom belong to the local branch of the cooperative society—who will have to vote on whether they want to buy the machine or not.

As I was saying, Comrade Chipkoff showed up very well. Unfortunately, however, the tractor did not, and we had a slight accident. As we were driving around a small pasture lot on the outskirts of the village, I was suddenly startled to see the motor burst into flames. Comrade Chipkoff, who was driving, at once stopped the machine. We leaped to the ground and began throwing handfuls of dirt onto the fire, and we managed to extinguish it before it did any damage.

Examination showed that the newly designed gasoline tank had been resting on the newly designed support bracket in such a way that the natural vibration of the motor had caused said bracket to wear a small hole in said tank. The gasoline had poured out, and had in some way caught fire, probably from the hot exhaust manifold.

Fortunately, there was very little fuel in the tank at the time, so the fire was small. And the only repair work needed was the temporary plugging of the leak with cotton and shellac. But you can understand that I was greatly annoyed at having to apologize to Mr. Krimsky, and to explain all over again that this was a new model and there were a few things that were not yet quite adjusted.

Note: I would respectfully suggest that you ask the engineering department why in the name of heaven, after all these years of building tractors, they have no more sense than to design a new model with a gasoline tank placed in such a way that it cannot help but wear a hole in itself.

One good thing about this accident was that it demonstrated Comrade Chipkoff's presence of mind and efficiency in an emergency. I have every reason to believe that he will acquit himself nobly in our great plowing demonstration tomorrow.

It is now after supper. In a few minutes we are going down to a big village meeting at the communist clubhouse. An official from Moscow will be present to tell the people about the plans of the Soviet Government for agricultural improvement, and Mr. Krimsky will explain to them all about the tractor. After the meeting, we will have a good night's rest, and bright and early tomorrow morning we will show them what we can do.

Yours,
ALEXANDER BOTTS.

ALEXANDER BOTTS
EUROPEAN REPRESENTATIVE
FOR THE
EARTHWORM TRACTOR

ODESSA, UKRAINIA, U.S.S.R.
WEDNESDAY, JUNE 6, 1928.

MR. GILBERT HENDERSON,
EARTHWORM TRACTOR COMPANY,
EARTHWORM City, ILLINOIS.

DEAR HENDERSON: It is with a feeling of deep despondency that I write this report. Our entire demonstration—mentally, morally, physically and mechanically—has gone completely haywire. It is hard for me to decide whether I am more disgusted with the absurd proceedings of the Soviet Government or with the pathetic performance of our boasted new-model thirty-horsepower Earthworm tractor. Both of them, to my way of thinking, are completely lousy.

The trouble began at the village meeting last night. Gadget and I, of course, could not understand any of the talk, but we could see that there was a very hot discussion going on. At the end of the meeting, Mr. Krimsky told us what it was all about. It appeared that the visiting Bolshevik from Moscow, with the support of a certain element in the town, had introduced, and finally carried, a resolution to expel our good friend Comrade Chipkoff from the local branch of the cooperative association. This was done on the ground that he was a *kulak*.

"And what," I asked, "is a *kulak*?"

"He is a man," explained Mr. Krimsky, "who violates the fundamental principles of communism by exploiting the labor of others for his own selfish gain."

"A sort of a grafter or profiteer?"

"Exactly. A *kulak* is a dishonest men."

"But Comrade Chipkoff seems to be such a good egg," I said. "I can't believe that he is really dishonest."

"I am sorry to say that he is. It was pretty definitely proved that he has been hiring laborers to help him with his farm work, by which he has made a selfish profit."

"You mean he didn't pay his hired help?"

"Oh, yes, he paid them very good wages. His crime was due to the fact that he made a profit for himself."

"But that isn't any crime at all back home," I said.

"Well, it is here," said Mr. Krimsky, "so poor old Chipkoff has been expelled from the association, and he is no longer eligible to drive the tractor. They have elected another man for this job."

"I suppose there is nothing we can do about it," I said, "but it certainly is a shame. Comrade Chipkoff has the makings of a swell tractor operator. What sort of a lad is our new candidate?"

"He is a *byedniak*."

"What kind of an animal is that?"

"It is not an animal. A *byedniak* is one of the poorer class of peasants. And the particular *byedniak* who has been chosen to drive the tractor happens to be Comrade Ivan Tschukovsky."

"Not that half-witted orangutan that stole the carburetor?"

"It was Comrade Tschukovsky that removed the carburetor," admitted Mr. Krimsky. "But it appears that he is a very good communist, and has a great deal of influence with the other *byedniaks*, and also with the *seredniaks*."

"Pardon me?"

"A *seredniak* is one of the middle class of peasants."

"You certainly have all these bozos ticketed," I said. "But this whole proceeding sounds awful funny to me. I very much doubt if we want this Comrade Tschukovsky for an operator."

"He is just the man for us," said Mr. Krimsky. "He has been very active against the *kulaks*, and he has brought so many malefactors to justice that they have given him the name of Ivan the Terrible. He is very popular with the majority of the villagers—particularly, of course, the poorer elements. He has great influence. He can be of great assistance to us in making a sale."

"Then I suppose we shall have to use him," I said. "But I still have my doubts. He may have influence, and he may be a rip-snorting *kulak* bouncer, but that does not mean he would be so hot as a tractor driver."

"Didn't you tell me that this new model was so simple to drive that a mere child could handle it?"

"I guess you win," I said. "I believe I did say that. We will consider the matter settled. And now let's go back and get some sleep."

"All right," said Mr. Krimsky. "And that reminds me—we can't spend the night with Comrade Chipkoff after all."

"Why not?"

"He is now in disgrace, and it would be very unwise for us to associate with him. I have arranged for us to stay with Comrade Ivan Tschukovsky."

"OK," I said. "Let's go."

We got our suitcases from the Chipkoff house and took them over to our new residence. And here Gadget and I got the big shock of the evening. Ivan's house was even more terrible than he was. It was a one-room log shack with a dirt floor and a tremendous brick stove, the top of which was used as a bed by Ivan, his wife and four children. They assigned us a pile of straw in one corner. In another corner was a small pig—fortunately, penned in by some old planks—and there were about a dozen hens which roosted on the table and on the two rough wooden benches. We tried to be good sports, and settled ourselves on the straw, without, of course, removing any of our clothes. But before long we were attacked by a small army of some sort of minute creeping fauna. So Gadget and I promptly

"They assigned us a pile of straw in one corner."

moved outside. And soon afterward Mr. Krimsky followed. Luckily, it was a warm night and we managed to get a small amount of sleep.

At the first crack of dawn, we got hold of old Ivan the Terrible and started to give him a lesson in driving the tractor. And from then on, our troubles increased very rapidly. Although Mr. Krimsky, acting as interpreter, explained everything to him with the greatest care, the poor thick-witted baboon did not seem able to grasp even the first principles. After an hour's work he seemed to know even less than when he started. Accordingly, we adjourned for breakfast, and after a miserable meal served by Ivan's greasy wife, we discovered that all the villagers had assembled and it was time to put on the great plowing demonstration. They had chosen a field to the east of town, on the edge of the cliffs overlooking the River Bug.

I decided to start things off myself. I climbed into the driver's seat and had Comrade Ivan Tschukovsky sit beside me. I hooked onto the big gang plow and drove out to the field, followed by several hundred of the villagers. I then struck a back furrow from the side of the field nearest the village clear over to the edge of the bluff, leaving only a narrow headland to be plowed later.

As long as I drove the machine myself, everything went beautifully, and for a few brief minutes I had a feeling that all would be well. The soil was a wonderful, rich, black loam which turned over so easily that

I was able to pull the entire six bottoms, set right down to a depth of fourteen inches. The motor roared, the machine rolled smoothly back and forth over the field, and I could see that the villagers were gazing at my performance with the deepest admiration and awe.

After four or five rounds, I stopped and consulted with Mr. Krimsky and Gadget, and Mr. Krimsky said it was now time to let old Ivan have a chance at the controls. I had him take the driver's seat and I sat beside him. We started off. We followed the last furrow which I had made across the field, and at first we got along fairly well. But when we reached the far end, I was startled to observe that Ivan made no effort either to stop the machine or to turn it around. The stupid brute merely sat perfectly still with his tremendous hands tightly grasping the steering wheel.

"It is time to turn!" I yelled.

Ivan did nothing.

"Turn it around!" I yelled.

Still Ivan did nothing I grabbed the wheel myself, but Ivan was such a powerful creature, and he was holding it so rigidly, that I could not do a thing. And all this time we were rapidly approaching the edge of the bluff which overlooked the Bug River.

"You poor sap!" I yelled. "Turn that wheel!"

At last the huge gorilla seemed to get the idea. He gave one terrific heave with his massive arms. And the wheel, the steering post, and the entire precision-movement, super-delicate steering mechanism came out by the roots. Ivan twisted the fragments hither and thither in a vain effort to steer. But it was no use; everything was completely disconnected. And the tractor was almost at the edge of the bluff. I reached over and cut off the ignition. With my other hand, I jammed on the emergency brake, which I had readjusted so that it took hold all right. The machine stopped on the very brink of the bluff, with the front end of it hanging out in space.

For a moment, I thought we were safe, but I was mistaken. I suddenly felt that the edge of the bluff was beginning to crumble beneath the weight of the tractor. Apparently Ivan had enough sense to know what was going on. With a clumsy, apelike bound, he jumped out of one side, while I leaped gracefully out the other. And the next instant my beautiful new-model thirty-horsepower Earthworm, with all its artistic red paint and everything, disappeared.

I stuck my head over the edge and looked down. I saw the great machine rolling and bounding across the sloping ground at the foot of the almost perpendicular cliff. Finally it came to rest on the level bottom land, a couple of hundred feet below me. And then, for the second time in two days, it burst into flames. This time there was no chance to throw on dirt. And there was a full tank of fifty gallons of gasoline. Obviously there was nothing to do but let it burn. The poor machine was such a complete wreck that I didn't even take the trouble to climb down the cliff to look at it.

The villagers all came rushing up in a high state of excitement, but I paid no attention to them. All at once I got mad. My disposition had not been improved by the uncomfortable night I had spent, and this disaster was more than I could stand. I walked right up to Mr. Krimsky and told him exactly what I thought of him and of the Soviet Government. After a certain amount of more or less general derogatory remarks, which I will not repeat, I ended up as follows:

"This is all the fault of your silly communistic theories. You want to throw out all the good, intelligent, efficient farmers like Comrade Chipkoff, and you insist on turning over the important jobs to mental defectives like this Ivan the Terrible. I should think when you are getting up a new system of government you'd get up one that has some sense to it."

About this time Mr. Krimsky himself began to get a little sore. "You are all wrong," he said. "It was not Ivan's fault; he tried to turn the machine around. The trouble is all with your boasted super-delicate steering mechanism. 'One touch from a baby's hand and the whole machine turns around.' Indeed! One touch from anybody else's hand and it comes to pieces. I should think that when you people build a new model you would get out one that had some sense to it."

Note: I will have to admit that there was a little justice in what Mr. Krimsky said. I would suggest that you tell the engineering department that there is such a thing as making a tractor a little too super-delicate. But, of course, I wouldn't admit anything like that to Mr. Krimsky.

"I do not care to argue with you," I said in a very loud voice, while all the villagers gathered around. "I only want to tell you that your whole Bolshevik system of government here is rotten, and it would be better for the country if you had some sort of a Czar back again who would run the country properly."

"You had better be careful," snarled Mr. Krimsky. "There may be agents of the Gay-Pay-Oo in this crowd, and some of them may understand English."

"And what," I asked, "is the Gay-Pay-Oo?"

"It is the secret police who are charged with suppressing just such counterrevolutionary sentiments as you have expressed. It may interest you to know that they would have the power to send you to Siberia, or to the dreadful prison on Solovyetzky Island."

"I care not that," I replied, snapping my fingers with a magnificent gesture, "for your old Google-Goo-Gay, or whatever it is. I only hope they are here and can understand what I say. My wife and I are completely through with you and your so-called government. We are going back to Odessa at once, and from there we are sailing on the first boat to sunny Italy. Goodbye."

I took Gadget by the arm and we marched back to the village and down to the railroad station. Fortunately, there was a train for Odessa in just a few minutes, and we got aboard. I noticed that Mr. Krimsky got on the same train, but he had sense enough to ride in another car. The trip to Odessa was uneventful, except for the fact that my angry passions began to subside.

As you know, I am naturally of a pleasant, friendly disposition, and I do not stay mad at anybody for very long. Consequently, by the time we reached Odessa. I had regained my usual poise and had decided that, after all, there is no sense in nourishing grudges against people—particularly people like Mr. Krimsky, who is, at bottom, a fairly decent chap.

Apparently Mr. Krimsky had the same idea. Gadget and I met him on the station platform and we all shook hands most cordially and apologized for our lack of courtesy earlier in the afternoon. Mr. Krimsky suggested that we might still be able to do business together. I agreed. And sometime later in the week we are going to put on a demonstration with the old-model sixty-horsepower tractor. In a future report I will let you know how we come out.

As we started for the hotel, Mr. Krimsky said: "I have the greatest admiration for you tractor people and for the way you are developing a new model with so many novel and improved ideas. You are public benefactors. Keep up the good work. But, whatever you do, don't try to sell me your new model until you have it working a lot better than it is now."

"Well spoken," I said. "And my sentiments toward you are exactly the same."

"How so?"

"I have the greatest admiration," I said, "for you Bolsheviks and for the way you are developing a new system of government with so many novel and improved ideas. You are public benefactors. Keep up the good work.

But, whatever you do, don't try to sell me your new ideas until you have them working a whole lot better than they are now."

"That also was well spoken," said Mr. Krimsky.

"Of course it was," I replied. "And before I wish you good evening I have one more message. I want to present you with the remains of that poor old tractor out there on the banks of the River Bug. I hope you will accept it, with the compliments of the Earthworm Tractor Company, as our contribution to National Junk Week."

And that is all at present from your hard-working salesman.

ALEXANDER BOTTS.

TECHNICAL STUFF

ILLUSTRATED BY TONY SARG

ALEXANDER BOTTS
EUROPEAN REPRESENTATIVE
FOR THE
EARTHWORM TRACTOR

HOTEL LONDONSKAYA,
ODESSA, UKRAINIA, U.S.S.R.
THURSDAY, JUNE 7, 1928.

MR. GILBERT HENDERSON,
EARTHWORM TRACTOR COMPANY,
EARTHWORM CITY, ILLINOIS.

DEAR HENDERSON: In case you did not receive my report of yesterday, I will explain that the first three days of my great selling campaign—by which I hope eventually to unload thousands of good American Earthworm tractors onto the sturdy peasants of Ukrainia—were not an absolutely complete success. We landed at Odessa on Monday. On Tuesday we arrived with the new-model thirty-horsepower tractor at the village of Usk, on the banks of the River Bug, about a hundred and fifty *versts* from Odessa. On Wednesday we held a plowing demonstration in the hope that we could sell the tractor to the Usk Cooperative Association. But, owing to the stupidity of Comrade Ivan Tschukovsky, who was driving, the tractor ran over a cliff and was completely destroyed. Last night, following this disastrous incident, I returned to Odessa.

Do not get the idea that I am discouraged. Many people, of course, would have been so overwhelmed by this disaster that they would have given way to despair. But troubles such as these only spur me on to greater efforts. And this morning, bright and early, I got hold of Mr. Krimsky. This gentleman—as I explained in my former report—is the wonderfully efficient and highly educated executive of the Union of Ukrainian Cooperative Societies.

"Mr. Krimsky," I said, "though it is true that we have lost one of our tractors, we are not yet licked. We still have the old-model sixty-horsepower machine in the warehouse here at Odessa. I propose that we ship it out to the village of Usk today, that we go out ourselves tomorrow and that we make a second demonstration."

"That will be all right with me," said Mr. Krimsky.

"This time I will do all the driving myself, so there will be no mishaps," I continued.

"It is a good idea."

"Furthermore, I would suggest that we ride along in the freight car with the tractor. It is still National Junk Week, and it is possible that if the tractor is unguarded, somebody else may come along and start stealing various parts off it."

(In my previous report I explained that National Junk Week, which is now going on, is a time when all good citizens are urged to bring to certain designated depots all the scrap iron, old brass and similar refuse that they can find. All this stuff, which otherwise might go to waste, will supply needed raw materials to the factories of the Soviet Union. The inhabitants have been so impressed with National Junk Week that they have been gathering material everywhere, and when we shipped the other tractor, an overzealous patriot swiped the carburetor and the magneto, which he intended using as his contribution to the cause. Fortunately, we recovered the missing parts, but we might not be so lucky the next time.)

Mr. Krimsky agreed with me that it might be wise to ride along with the tractor and guard it. Accordingly he departed for the railway freight terminal to make the necessary arrangements. After agreeing to meet him there late in the afternoon, I returned to the hotel and explained the arrangements to Gadget. Gadget, I regret to state, was inclined to be just a trifle peevish.

"Of course," she said, in a faintly sarcastic tone of voice, "I always try to be a good and helpful wife to you, Alexander."

"You certainly are," I said. "Throughout our entire European tour your services as interpreter and as business adviser and assistant have been invaluable."

"Well, I would just as soon act as your interpreter in France and Italy, where I understand the languages. But here in Russia I can't talk the lingo any more than you can. Mr. Krimsky understands both English and Russian; he is going along with you; so you don't need me. I have a good mind to stay right here in Odessa where it is clean and comfortable and civilized."

"You mean you don't want to come with us?"

"Oh, I'll probably go with you, if you insist," said Gadget, "but it does seem to me that you are dragging me around to a lot of awful funny places. That house belonging to Comrade Tschukovsky—where we were supposed to stay on our last trip to Usk—was one of the filthiest holes I have ever seen. If I wanted to be perfectly accurate, I might even describe it as lousy."

"Perhaps," I replied, "that was to be expected, considering the place was so near the Bug River." This was supposed to be a wisecrack, but it didn't get over very well.

"I am sure of one thing," said Gadget. "I'll never enter that house again."

"You won't have to," I said. "We can sleep outdoors."

"This certainly is a wonderful trip you are offering me. We go there in a freight car and we sleep outdoors."

"All right," I said, "if you don't want to go, you don't have to. It makes absolutely no difference to me whether you do or not."

"In that case," said Gadget, who seemed to be in a somewhat contrary frame of mind, "I think I will go."

And so it was decided. It is now four o'clock in the afternoon, and we are all ready to start for the freight terminal. In my next report I will tell you how everything goes.

Very sincerely,
ALEXANDER BOTTS.

ALEXANDER BOTTS
EUROPEAN REPRESENTATIVE
FOR THE
EARTHWORM TRACTOR

AT THE FREIGHT TERMINAL, ODESSA.
THURSDAY, JUNE 7, 1928, 7 P.M.

MR. GILBERT HENDERSON,
EARTHWORM TRACTOR COMPANY,
EARTHWORM CITY, ILLINOIS.

DEAR HENDERSON: I have run into a very curious little series of difficulties, which for a time threatened to disrupt our plans, but which I have solved in a very interesting and very original way.

When Gadget and I arrived at the freight terminal, Mr. Krimsky met us and reported that the car was ready, and that he had received permission for himself and for me to ride alone with the tractor. Unfortunately, however, it was against the rules of the railroad for a woman to ride on a freight car.

"What a silly rule," I said.

"Not at all," said Gadget. "It settles everything very nicely. I didn't want to go on this idiotic expedition anyway. I will stay in Odessa."

"I hate to leave you behind," I said.

"Don't worry about me. I will spend the rest of the afternoon doing a little shopping, and then go back to the hotel and have a good supper and a comfortable night's rest. And tomorrow I may do a little sightseeing. Goodbye and have a good time."

With these words Gadget left us and went back uptown. I got busy and cranked up the old sixty-horsepower tractor, drove it across the platform and loaded it on the boxcar which was waiting. Mr. Krimsky and I then sat down to wait for the train to start.

Before long a large bozo in uniform came walking by and asked to see our papers. Every time you move over here, and every time you stop anywhere, you have to show your papers to some official and go through a lot of red tape. We produced our passports and the permission of sojourn, with all the funny-looking photographs and official stamps and visas and what not. After looking them over, the man in uniform had a long discussion with Mr. Krimsky. Then Mr. Krimsky explained matters to me.

Apparently our friend was some sort of a police officer whose job it was to see that no funny business was pulled off around the freight yards. And he was so surprised to discover two such good-looking chaps as Mr. Krimsky and myself preparing to take a ride in a freight car that he decided he ought to investigate our papers very thoroughly. Now, it happened that Gadget and I had a joint passport, made out in the name of Alexander Botts and wife. We also had a joint permission of sojourn, which is a paper they give you on your arrival and which yon have to have in order to remain anywhere inside the U.S.S.R. (This stands for Union of Socialist Soviet Republics.)

The cop looked at my papers and noted that they called for a man and a wife. And when he failed to discover the wife anywhere around, his suspicions increased to a positive certainty that I was engaged in some very dark and criminal proceedings. Mr. Krimsky explained everything to him very carefully, but he was both dumb and stubborn, and he insisted that I could not travel alone on papers which called for a man and a wife. He insisted that we come to the central office with him. Mr. Krimsky told him he was crazy and that he would make us lose our train, which was due to leave within an hour. But he remained completely bullheaded, and we had to go along with him.

First we visited some sort of a police station, from which we were sent along to the People's Commissariat for Foreign Affairs, or some such thing. This place was on the Uliza Pushkinskaya. Mr. Krimsky had a long and angry discussion with the man who seemed to be in charge of the office. Of course I couldn't understand anything that was said, but I had a feeling that old Krimsky was doing a lot of high-powered arguing which was not getting him anywhere.

"This man," he said, "is incredibly stupid."

"Practically all of them are," I said.

"Practically all who?" he asked. "Are you trying to insinuate that practically all Russians are stupid?"

"No," I replied. "Practically all officials. I have had arguments with customs officials in Italy, sergeants of the military police in the A.E.F., traffic cops in Illinois, and many others. And they are all just as stupid as you say this bird is."

Mr. Krimsky smiled somewhat bitterly. "Yes, yes," he said. "I see what you mean, and I agree with you—except that this bird, as you call him, is obviously gifted with a deeper and far more solid stupidity than anybody you have met up to this time."

"Mr. Krimsky told him he was crazy and that he would make us lose our train, which was due to leave within an hour. But he remained completely bullheaded, and we had to go along with him."

"What does he say?" I asked.

"In the first place, he says the policeman is in error; it is perfectly legal for you to travel alone on papers made out for you and your wife."

"Then we are all right," I said.

"Apparently not. If you want to leave town, you will have to take your papers with you. If you take them, Mrs. Botts will be left without any. And in that case this officious fool says it will be his duty to send an officer to the hotel to arrest Mrs. Botts for living in Odessa without the proper permission of sojourn. It looks as if you and your wife—as long as you have but one set of papers—will have to remain together."

"But that is absurd," I said. "I have to go to Usk, and I have to go this afternoon on that freight train. How can Gadget go along with me? She is out shopping somewhere, and we couldn't find her before train time. And even if we could, the railroad company wouldn't let her ride on the freight train. And besides, she doesn't want to go to Usk. You tell your friend here that there is no sense to what he says."

"I have already told him, but it does no good."

"Ask him if he can't issue us separate papers."

"Maybe he can," said Mr. Krimsky. He had another long argument with the half-witted official, and then turned once more to me. "He says that you both entered the country on a joint passport. Therefore, he can do nothing. Personally, I don't think he knows what he's talking about. I very much doubt if there are any such rules and regulations as he claims to be following. I could take the matter up with his superiors in Moscow, but we couldn't get an answer for several days. And in the meantime we seem to be stuck. He insists that as long as your status remains unchanged, it would be illegal for him to issue you separate papers."

"What does he mean by our status remaining unchanged?"

"I'll ask him." After another conference with the official, Mr. Krimsky explained. "He says that if one of you died, he would be forced to issue revised papers to the survivor at once. And he would have to give you separate papers in case you were divorced."

"That doesn't help as much," I said. And then I had one of those brilliant inspirations which once in a while seem to come into active minds such as my own. "Wait a minute," I said. "Haven't I heard that a divorce is very easy to get in this country?"

"As a matter of fact," said Mr. Krimsky, "the marriage and divorce laws of the Soviet Union are the most progressive and the most liberal in the world. They are based on the high principle that love and marriage are too sacred to be hampered and stultified by absurd and deadening restrictions. In the communist state, if a man wishes a divorce, all he has to do is to apply for it at the government marriage and divorce bureau, and it is given him at once."

"This applies to foreigners also?"

"Certainly."

"What do you mean by 'at once'? How long does it take?"

"It might take five minutes."

"And if a man got a divorce, could he remarry anytime he wanted to?"

"He could."

"He could marry his divorced wife five minutes later? Or the next day? Or anytime at all?"

"Absolutely."

"Where is this divorce office?" I asked.

"But surely," said Mr. Krimsky, "yon aren't actually thinking of divorcing your excellent wife?"

"I certainly am."

"But I can't believe it," said Mr. Krimsky, in a shocked tone of voice. "I hope you will pardon me for interfering in your private affairs, but Mrs. Botts is so charming, so accomplished, and you are so obviously suited to each other that you would be a complete fool if you tried to get rid of her."

"Exactly so," I agreed. "I am not trying to get rid of her. We'll get married again as soon as I come back from Usk. It will be merely a technical divorce, so I can get the papers necessary for this trip. If we hurry, we'll just have time before that freight train leaves. Let's go."

"I don't like this at all," said Mr. Krimsky "Our divorce laws are liberal, and it is proper that they should be so. But they should not be used for trivial purposes."

"Well, it's not my fault," I said. "It's the fault of the government over here, which makes such trivial and frivolous laws about passports."

"I tell you I don't believe there are any laws such as this man pretends to quote."

"All right," I said. "It's the fault of the government for employing half-witted officials that know so many laws that aren't so. If these officials insist on acting foolish, I'll show them that I can act just as foolish myself. I'll beat them at their own game, and we'll make this trip in spite of all their technicalities. Let's go."

Mr. Krimsky was still rather doubtful, but he finally agreed and took me over to the marriage-and-divorce-registration bureau, where the man in charge made out the necessary papers with surprising promptness. As soon as I had signed the various copies and shelled out the necessary kopecks for the fee, we hurried back to the People's Commissariat for Foreign Affairs, where our feeble-minded friend made out separate permissions of sojourn for Gadget and myself.

We then took a cab and rushed around to the Hotel Londonskaya. Gadget had not yet returned, so we left her certificate of divorce and her permission of sojourn with the proprietor. Fortunately, he spoke a little English. We told him to explain to her that this was a technical divorce, made necessary by the laws regarding passports and permits. We then tore down to the freight station and found that we were in plenty of time for the train. In fact, we have been here more than an hour, and I have been improving the time by writing this letter. Apparently, the train will move out in about five or ten minutes, so I will have time to mail it before we leave.

In conclusion, I may say that I really feel rather proud of myself. It takes an ingenious American like me to figure out clever ways of getting around governmental red tape. It is probably due to my ability along these lines that I get along so smoothly and so pleasantly everywhere I go. Every time I think of this divorce business, I have to chuckle inwardly. I can hardly wait to see Gadget again. She certainly will have a good laugh when she sees those papers.

With best wishes from your—technically speaking—bachelor salesman,

ALEXANDER BOTTS.

ALEXANDER BOTTS
EUROPEAN REPRESENTATIVE
FOR THE
EARTHWORM TRACTOR

USK, UKRAINIA, U.S.S.R.
FRIDAY AFTERNOON, JUNE 8, 1928.

MR. GILBERT HENDERSON,
EARTHWORM TRACTOR COMPANY,
EARTHWORM CITY, ILLINOIS.

DEAR HENDERSON: It gives me great pleasure to report that we have held our demonstration, and all is well, except for one thing, which I will refer to later.

Our ride on the freight train was uneventful and reasonably comfortable. This morning, on our arrival in the village of Usk, we heard very good news to the effect that Comrade Ivan Tschukovsky was no longer eligible to act as tractor operator. I believe I have explained to you that this man had been elected at the village meeting to the job of tractor operator, and during our first demonstration had been so clumsy as to let the tractor roll over a cliff and smash itself to pieces. And I had determined, as I stated, to fire old Ivan the Terrible myself, even if I insulted the whole village by doing it. But, fortunately, this decision was taken out of my hands.

It appeared that last night Ivan had been caught selling *samogon* and was safely lodged in the Usk jail. *Samogon*, I might explain, is a very powerful variety of bootleg white mule. They don't have prohibition in this country, but the government has a monopoly on the liquor business and makes a lot of money out of the sale of vodka. They are, therefore, fairly severe with bootleggers who sell the homemade *samogon*, which apparently is produced in considerable quantities in small stills all over the country. And poor old Ivan the Terrible, much to my relief, had fallen foul of the law.

As none of the local boys had yet been designated to take Ivan's place as operator, I hooked onto the plow myself and spent the entire morning plowing the large field between the village and the Bug River. Everything went fine and the villagers seemed highly enthusiastic. In the afternoon we picked out three or four intelligent young men from among the inhabitants and gave them driving lessons. I was very careful to keep them away from the neighborhood of the bluff overlooking the river. I didn't want any more accidents. But this precaution was probably unnecessary, as all these young men were competent and learned very rapidly. These people here have had very little experience with machinery, but they have as much brains as anybody, and there is no doubt but what they can learn to operate tractors successfully.

Tonight there will be a village meeting, at which they will decide definitely whether or not they want to buy the tractor. I had intended to stay over until tomorrow at least, but I have been forced to change my plans on account of the difficulty which I mentioned at the beginning of this letter.

About the middle of the afternoon, Mr. Krimsky received a telegram from the proprietor of the Hotel Londonskaya in Odessa. This telegram was in Russian. Translated, it read as follows:

THE AMERICAN LADY IS MAKING US MUCH TROUBLE WE THINK SHE HAS GONE INSANE PLEASE COME AT ONCE AND ADVISE US WHAT TO DO WITH HER

Naturally, this wire has disturbed me very much. Apparently it refers to Gadget. As far as we know, she is the only American woman at the Hotel Londonskaya. I cannot imagine what has happened. Gadget has always been perfectly normal and well-balanced, and I can't think of anything that could have disturbed her or worried her in any way.

Probably, the proprietor has made some mistake. It may be some other woman. But I am taking no chances; I am leaving Mr. Krimsky in charge here, and I am hurrying back to Odessa on the afternoon train, which leaves in about half an hour.

I am really very much worried.

Yours,
ALEXANDER BOTTS.

ALEXANDER BOTTS
EUROPEAN REPRESENTATIVE
FOR THE
EARTHWORM TRACTOR

HOTEL LONDONSKAYA,
ODESSA, U.S.S.R.
FRIDAY EVENING, JUNE 8, 1928.

MR. GILBERT HENDERSON,
EARTHWORM TRACTOR COMPANY,
EARTHWORM CITY, ILLINOIS.

DEAR HENDERSON: I have been having a terrible time. I am practically out of my mind. I have engaged passage on an Italian steamer leaving here tomorrow noon for Trieste, Italy, and I am shaking the black earth of Ukrainia from my feet forever and ever. How I hate this place. I haven't heard whether or not the inhabitants of Usk have decided to buy the tractor. And what is more, I don't care. I don't care if I never sell another tractor.

In view of the fact, however, that you are paying my salary and expenses, it is only reasonable that I should let you know the causes of my sudden complete demoralization. I will, therefore, give you a brief sketch of the events of the evening. I am so upset that I can't remember for sure, but I think I wrote you this afternoon of the telegram which called me back from Usk. I arrived in Odessa this evening and came at once to the Hotel Londonskaya. I was starting up to the room to see Gadget, when the proprietor stopped me.

"She has given orders," he said, "that she will see no one."

"Who will see no one?" I asked.

"Madame Botts, your former wife."

"Former wife? What do you mean—former?"

"I thought you were divorced."

"Technically, yes, but actually, no. She is just as much my wife as she ever was, and I'm going up to see her."

"But she left word she would not see you."

"I don't care if she did. I am going up anyway."

"But it will do no good. She has locked the door and will open it to no one."

"All right," I said. "I seem to remember that there is an ax hanging up in the hall beside the fire extinguisher. I will break open the door."

"I beg of you, Comrade Botts, to control yourself. Look, here is a letter which she wrote and which she commanded me to give you in case you came."

"Give me that letter."

I opened it. I read it. As I never discuss my private affairs with business associates, I will not quote this letter in full. However, as I said before, I feel it is my duty to tender you an explanation as to why I have so completely lost interest in the tractor business. And for this reason I will give you a few pertinent paragraphs which will show you what a mess I am in. The letter read, in part, as follows:

> I had always supposed that you were a gentleman and a man of honor. I never would have married you in the first place if I had supposed you were capable of anything so mean, so underhanded, so lowdown and so completely and ineffably execrable. I would not believe it now, except for the fact that I have seen your name as you wrote it on the papers. It is your signature, all right. I know it so well that there could be no mistake about that.
>
> Very well. If you are so fickle and vacillating that you have wearied of me; if you have found some younger and more attractive hussy among these immoral Russians, or elsewhere; if you have become completely corrupted by your association with these conscious-less Bolsheviks and have sunk so low that you are willing to employ their wicked and shameless divorce laws—I am not the one to stand in your way. From now on, I am as completely through with you as you are with me. I will not seek to hold you back. In fact, I will even go so far as to wish you success; may you have as much happiness as is possible in your new and shabby existence. I want you to understand, however, that if you should ever come to your senses—as, doubtless, you will—and desire to come back to me, I will never for an instant have anything to do with you.
>
> I am leaving tomorrow for Trieste, Italy, on the Italian boat, *Santa Maria*, which sails from the south side of the Platonovski Mole. From Trieste, I am going straight back to my home in California, where I am appreciated and where I will take up the old free life which I so unadvisedly abandoned at the time of our marriage. Goodbye forever.

After reading this letter two or three times, I came to the conclusion that Gadget was actually a bit sore at me on account of this divorce business. And I will admit that I was in something of a quandary as to just what I ought to do about it. I glared about the hotel lobby and noticed that the proprietor and a number of the bellhops and porters were eyeing me somewhat apprehensively.

"Come here, you!" I said. "I want to talk to you!" The proprietor stepped up. "I'll bet you are the guy," I continued, "that got me in wrong around here. Just what did you tell my wife?"

"You mean your former wife?"

"Shut up! Don't feed me any more of that 'former' stuff. What did you tell my wife?"

"Just what you told me to, sir."

"And exactly what was it that you think I told you and that you passed along to her?"

"I gave her the papers," he said, "and told her that you had left them. She wanted to know what they were, so I translated them for her, and explained that you had decided to leave her, so you had got a divorce and arranged to go out of town on the evening train."

"Is that all you told her?"

"As far as I remember, it was. What more could I have said? The case, as I understood it, was simple. You wished to leave your wife, so you got the divorce and left."

"You poor sap," I said. "You bothering, clobbering idiot! I told you to tell her that the divorce was only technical, and that I would marry her again when I got back from this short trip. Don't you remember that I told you that? Didn't you understand that this divorce was only for convenience in traveling?"

"No, I didn't understand any such thing. And I don't understand it yet. It does not make sense. Why should you divorce your wife if you had already decided to get married to her again in a day or two?"

"Shut up," I said. "I have no time to explain all this complicated business to you. Besides, it is too late now, and you couldn't understand it anyway. You are too dumb. What did my wife do after you gave her the papers?"

"She did plenty. She told me that I was a liar and a slanderer, and then she lost control of herself completely. She seems to be a very high-spirited woman."

"Just what did she do?"

"Well," said the proprietor, "she slapped me in the face."

"Served you right," I said. "And what happened after that?"

"She called me a liar several times more, and said she didn't believe it was true that you had got a divorce. So I got hold of an Englishman who is staying here at the hotel. He understands Russian and he went over the papers with her. Then, when she had examined your signature on them, she was finally convinced that I was telling the truth. She seemed to be much displeased. I would suggest that the next time you get a divorce from anybody, you break the news a little more gently. One should always be courteous and considerate in matters of this kind."

"Thank you," I said. "I don't want any wisecracking advice from you or anybody else. You say she was displeased?"

"More than displeased. Perhaps 'enraged' would convey the meaning better. She picked up a chair and threw it completely across the room. You can see how it smashed the window over there, which we have not yet had time to repair. Then she went up to her room and locked herself in."

"That was last night?"

"Yes."

"And she hasn't been out since?"

"Oh, yes. She went out this morning, and when she came back, she told me that she had engaged passage on the boat and that she would be leaving tomorrow. And before she went up to her room, she told me that she was going to keep her door locked, and that she didn't wish to see anyone. She has had her meals sent up to the room."

"Well," I said, "you certainly have got things balled up for me good and plenty. And now I suppose I will have to figure out some way of straightening them out again."

I sat down in the lobby and smoked a number of cigarettes while I turned the matter over in my mind. Finally I decided that it might be wiser and safer to send up a letter to Gadget before actually going to see her. As the hotel proprietor had remarked, Gadget is a high-spirited woman. And if she was in a state of mind which called for throwing chairs through panes of glass, I didn't want to find myself accidentally situated between her and a window.

Accordingly, I sat down at one of the hotel writing desks and composed a letter which ended something like this:

> This all a terrible mistake. And I am ready to admit that it was entirely due to my own clumsiness and stupidity in failing to make sure that you received accurate information as to exactly what I was

> doing. You see, in going to Usk I had to take all our papers along to show the police. If I had left you behind without papers, you would have been clapped in jail. I had to have papers. You had to have papers. And they wouldn't issue separate papers as long as we were married. So I got a purely technical divorce. It doesn't mean anything, except that now we will have to be technically married again. So everything is all right. I am waiting downstairs. I will come up as soon as you give the word. And then we can both have a good laugh together over this whole business.
>
> In conclusion, I wish to apologize once more for the stupid way I have handled things. It was all my fault, and I beg your forgiveness in a most abject spirit.

I sent my letter up by the porter. In about ten minutes he came back with a very short note from Gadget:

> Your explanation sounds awful fishy to me. Whoever heard of a technical divorce? I admit I am surprised that you have changed your mind again so soon. But it is too late now. You have given me the air once, and you won't get a chance to humiliate me by doing it again. I am leaving on the boat tomorrow. I never want to see you again.
>
> GADGET.

After reading this note, I rushed upstairs, in spite of the efforts of the proprietor to restrain me, and knocked and pounded on Gadget's door, but there was no response to my knock.

I knew she was there, so I made a long and earnest appeal for her to let me in. I explained everything in much greater detail than before. I apologized fully for the trouble I had caused and I begged forgiveness in what seemed to me a very heartrending manner.

Finally, there came a chilly response from inside the room. "It is no use," said Gadget. "I don't believe a word you say. You wanted to get rid of me, and you did. And now that you have changed your mind, I won't have you. And I don't care to discuss the matter. I am leaving on the boat tomorrow."

I then renewed my pleas and my apologies, but all to no purpose. I looked at the fire ax hanging on the wall, but decided that breaking down the door would only aggravate the situation. Finally, I went downstairs.

"I knocked and pounded on Gadget's door."

After pacing up and down the lobby for some time and smoking several packages of cigarettes, I went over to the steamship office and engaged passage on the same boat that Gadget was taking. It occurs to me that in the course of a long sea voyage I may be able to get Gadget to listen to me. And I may be able to convince her that, although my actions may have been peculiar, my intentions have been at all times of the best.

But I will have to admit that I feel very low in my mind. Gadget is such a strong-minded woman that it is perfectly possible she may actually refuse to have anything to do with me from now on. And in this case my whole future life would be ruined.

I have spent the last half of the evening writing this letter, and it is now midnight. I have engaged a single room for myself for the night, and I will now go to my lonesome bed and attempt to get some sleep.

Yours,
ALEXANDER BOTTS.

ALEXANDER BOTTS
EUROPEAN REPRESENTATIVE
FOR THE
EARTHWORM TRACTOR

ON BOARD THE S.S. *SANTA MARIA*.
SATURDAY, JUNE 9, 1928.

MR. GILBERT HENDERSON,
EARTHWORM TRACTOR COMPANY,
EARTHWORK CITY, ILLINOIS.

DEAR HENDERSON: Much has occurred since my last letter. Early this morning I went down to the People's Commissariat for Foreign Affairs and got my visa of departure. After this, I returned to the hotel, packed my few belongings and settled with the landlord. As I am always large-minded and liberal in money matters, and as this will go on my expense account anyway, I paid Gadget's bill as well as my own—including six rubles and forty kopecks for the busted window. Then I went down to the Platonovski Mole, arriving at half past eleven. The steamer was due to sail at twelve noon.

At 11:35 Mr. Krimsky suddenly appeared. He had called at the hotel, had been told that I was about to embark, and he had hurried down to see why I was leaving Russia so suddenly. I explained the situation to him and he was most sympathetic.

"Permit me to wish you good luck on your voyage," he said. "We would have enjoyed having you here longer, but you have done so well that I am sure we can now carry on without you. The Usk Cooperative Association has voted to buy the tractor, and the Ukrainian Association has authorized me to order five more of the same type of sixty-horsepower machines. Here is a draft of the Ustorg Corporation—our American agents—on New York, covering payment in full for the tractor we have bought. And here is the order for the five additional machines. You will note that they are to be shipped at once at the same price, C.I.F. Odessa."

"Thank you very much," I said. "I appreciate this more than I can tell you."

About this time Gadget arrived on the dock and walked past us on her way to the gangplank. Mr. Krimsky and I both wished her good morning most politely, but she strutted along with her nose slightly elevated and without returning our greeting in any way. We both followed her as she proceeded down the dock, with a couple of porters carrying her baggage. When she reached the gangplank, she was stopped by a group of officials. We saw her produce her papers. One of the officials looked them over, then shook his head and started a long harangue. Apparently, there was something wrong.

Mr. Krimsky and I walked to the gangplank and Mr. Krimsky asked the officials what it was all about. After a long and intricate conversation, he explained matters to me, while Gadget stood off at one side, pretending to pay no attention, but actually, of course, taking it all in.

"Mrs. Botts," he said, "cannot embark because she has no visa of departure. And she can't get one because she has no passport."

"She has that permission of sojourn," I said. "Why can't they stamp her visa on that?"

"It is against the rules. A foreigner who wishes to leave the U.S.S.R. must prove that he entered the country legally. To do this, he must present

"One of the officials looked them over, then shook his head and started a long harangue."

his national passport with the visa of entrance properly executed. The visa of departure is then stamped on the same passport."

"All right," I said. "Here is my passport, made out for both of us and duly stamped with the visa of entrance and the visa of departure."

Mr. Krimsky showed the passport to the officials, but they refused to accept it. "They say," explained Mr. Krimsky, "that this is all right for you alone, or for you and your wife. It does not apply to your divorced wife. She will have to have a separate passport."

"And how can she get one?"

"As there are no American consuls in Russia, she would probably have to send to Washington."

"In which case," I said, "she might not get it for months."

"Probably that is true," admitted Mr. Krimsky.

"What would she do in the meantime?"

"She would have to stay in Odessa. But that would bring up another complication. Her permission of sojourn is limited to thirty days. After it expired she could not remain in Odessa any longer."

"So then they would have to let her go home?"

"Oh, no. She would not be permitted to leave unless she had her passport."

"You mean she couldn't stay and she couldn't leave either?"

"That's it. No matter what she did, she would be violating the law and would be subject to arrest."

"In other words," I said, "her chances of staying out of jail would be very slim."

"Very slim, indeed," said Mr. Krimsky. He discussed matters a little further with the officials and then turned back to me. "There is only one solution," he said. "You will have to remarry your wife. Then you can both leave on your original passport."

I looked over at Gadget. "You heard what the gentleman said," I remarked. "How about it?"

"Absolutely not," said Gadget.

"I think," said Mr. Krimsky, "you had better reason with the lady a little. In the meantime, I will hurry uptown and bring down the registrar of marriages. I think I can get him here just in time to marry you before the boat sails; provided, of course, the lady decides to be sensible."

As Mr. Krimsky departed, there came a sudden deafening roar from the big whistle on the boat. It reverberated across the placid waters of the harbor and echoed back from oil tanks and grain elevators and from the monumental steps leading up into the town. Gadget and I both glanced at the clock on the pier. It was a quarter before twelve. The whistle was a signal that only fifteen minutes remained before the boat was to sail.

After the whistle ceased, there was a moment or two of silence. Then Gadget spoke up. "If you insist upon my going through this silly marriage ceremony," she said, "I suppose I shall have to do it. But kindly take note, Mr. Alexander Botts, that it will be a purely technical marriage—a marriage in name only. Never again will I live with you as your wife."

As I listened to these words, I will have to admit that my heart was filled with despair. But all at once a change came over me. I began to get mad. And at the same time I got one of my sudden inspirations. It occurred to me that I had accomplished absolutely nothing by all my attempts at courtesy and all my humble apologies and prayers for forgiveness, and that the situation demanded a complete change in tactic.

I regarded Gadget with a cold eye. And in gruff and brutal tones I proceeded to make a little speech.

"Gadget," I said, "I am completely sick of all this foolishness. I take back all my apologies. Everything I have done in this technical divorce business has been right. Everything you have done or said has been absolutely wrong. You have been acting like a nasty little spoiled child, and you know it. If I treated you as you deserve, I would cast you off forever and ever. But, out of pure generosity, I am going to give you one more chance. If you will admit that you have been acting like a fool, if you will humbly apologize for causing me so much trouble, if you will promise never to question any of my actions again, to love and to honor, I may possibly be able to do something for you. Otherwise, you may rot in this foreign land for the rest of your life. You have your choice, and it really makes very little difference to me which way you decide."

"Why, Alexander," Gadget said, "how could you be so cruel?"

"When it comes to cruelty," I said, "I am so far ahead of the Emperor Nero that there just is no comparison at all."

To these words Gadget made no response at all, so I occupied the next few minutes in strolling jauntily up and down.

After a while a horse-drawn hack came clattering out onto the pier, and Mr. Krimsky and the man from the marriage bureau jumped out.

"What has been decided?" asked Mr. Krimsky.

"Everything is fine," I said. "Apparently the lady has decided to remain behind, so I must be on my way. I am sorry to have brought this gentleman down here to no purpose, and I wish to thank you both for everything you have done. Goodbye!"

As I strode briskly toward the gangplank, the great whistle on the boat burst forth with another terrific shriek. It was 11:55. Only five minutes remained before the boat was to leave.

I showed my passport to the officials. They waved me onward and I started up the gangplank. By this time the whistle had relapsed into silence, and my ears were therefore able to detect a feeble and piteous cry from Gadget.

"Alexander," she said, "come back here!"

I stopped. "Did somebody call?" I asked.

"Yes," said Gadget. "I did."

"Well, well," I said in an extremely nonchalant tone of voice. "Is there anything I can do for you before I go?"

"Yes," she said. "I want you to marry me and take me with you."

"Anything to oblige a lady," I replied. "But I am afraid it is too late. I doubt if there is time before the boat sails."

"Please, Alexander," she said. "Don't desert me. Don't leave me behind. If you'll only take me with you, I'll never contradict you or argue with you again. I know I've been acting perfectly rotten. But I'm sorry. And I'll promise to be good all the rest of my life."

"Are you sure," I asked, "that this is honest repentance, and not just a dodge to get on the boat?"

"It is honest and complete repentance," said Gadget. "Please hurry or it will be too late."

Well," I said magnanimously, "as a great favor, I will take you with me. And I suppose the only honorable thing is to make an honest woman out of you by remarrying you."

I turned to Mr. Krimsky. "All right," I said, "bring on that marriage mechanic and let's get busy."

Mr. Krimsky and the official rushed up and we went through a short and, of course, completely unintelligible ceremony in Russian, with the various officials as witnesses. At the conclusion of the Russian part, I added a clause in English so that there would be no doubt as to certain important details.

"Do you, Gadget," I asked, "promise to love and honor me, Alexander, till death us do part?"

"I do," said Gadget.

"Fine," I said. I then kissed the bride, and it seemed to me that she rather liked it.

The two-minute whistle boomed out over the harbor and we rushed up the gangplank. And as the boat pulled out, we stood at the rail, waving cordially to Mr. Krimsky and the people on the dock.

"I must tell you, Alexander," said Gadget, "that I think you were really magnificent this morning. I was so proud of the manly way in which you stood up for your rights."

"Well," I said, "I will have to admit that I really wasn't so bad, at that."

Yours, with well-justified pride,
ALEXANDER BOTTS.

IN THE ENEMY'S COUNTRY

ILLUSTRATED BY TONY SARG

EARTHWORM TRACTOR COMPANY
EARTHWORM CITY, ILLINOIS
OFFICE OF THE SALES MANAGER

JUNE 10, 1928.

MR. ALEXANDER BOTTS,
TRIESTE, ITALY.

DEAR BOTTS: At a meeting yesterday of the board of directors of the company it was decided that you have not had sufficient success to justify us in keeping you in Europe any longer. This is not a criticism of your work. You have probably done everything humanly possible. But your sales have been few, and we have come to the conclusion that economic conditions are so unfavorable that we cannot hope to do a profitable business with Europe at the present time. We want you to get rid of all machines on hand—at reduced prices, if necessary. Then you and Mrs. Botts may come home as soon as convenient.

We have given similar orders to Mr. George McGinnis, whom we sent to Germany at the same time we sent you to Southern Europe, and who has so far failed to sell a single one of the five machines he took over with him. It has occurred to us that on your way home you might be able to arrange your plans so as to spend a few days in Germany with Mr. McGinnis. You have had somewhat better success in dealing with foreigners, and you could probably help him in selling off the tractors he has on hand. His address is Rhein Hotel, Coblenz.

Very sincerely,
GILBERT HENDERSON,
Sales Manager.

ALEXANDER BOTTS
EUROPEAN REPRESENTATIVE
FOR THE
EARTHWORM TRACTOR

HOTEL EXELSIOR,
TRIESTE, ITALY.
JUNE 24, 1928.

MR. GILBERT HENDERSON,
EARTHWORM TRACTOR COMPANY,
EARTHWORM CITY, ILLINOIS.

DEAR HENDERSON: Your letter of June tenth has been forwarded from Marseilles, and I got a good laugh over your account of the meeting of the board of directors. They must be getting weak-minded in their old age. Of course I have not sold many tractors yet. But I am preparing the ground. And within a month or two there will be so many orders pouring in that you wouldn't even dream of asking me to come home.

The second big laugh in your letter was the account of George McGinnis's efforts in Germany. Of course I never like to knock another man's ability, but it certainly is a lucky thing you thought of sending me to Germany to get things straightened out. There is a lot of work I could do here in Italy, but I feel it is my first duty to help out a fellow salesman in distress. Accordingly, Gadget and I are leaving at once for Coblenz, and we will let you know later how we get along.

Sincerely yours,
ALEXANDER BOTTS.

ALEXANDER BOTTS
EUROPEAN REPRESENTATIVE
FOR THE
EARTHWORM TRACTOR

RHEIN HOTEL, COBLENZ, GERMANY.
WEDNESDAY AFTERNOON, JUNE 27, 1928.

MR. GILBERT HENDERSON,
EARTHWORM TRACTOR COMPANY,
EARTHWORM CITY, ILLINOIS.

DEAR HENDERSON: Here we are. As soon as we arrived this morning, we had a conference with George McGinnis and heard all about the bad luck he has run into. He hasn't been able to do anything because of the violent prejudice against Americans. Apparently the feeling goes back to the old wartime hatred.

"The Germans," said George, "are a proud people. They can't forget that we defeated their armies and that we occupied part of their country for a time. They don't like Americans."

"I can't say I blame them," said Gadget. "If the Germans had captured and occupied my hometown in California, I'd probably hate them worse than rattlesnakes."

"Exactly so," said George. "And by the way," he went on, turning to me, "weren't you in the Army during the war?"

"Yes," I admitted. "And I was in the Army of Occupation after the war. In the spring of 1919 I was billeted right here in Coblenz with a very nice German family by the name of Pfeiffer. I want to call on them if I have time."

"I wouldn't do that," said George.

"Why not?"

"As long as you are working with me you don't want to let anybody know you fought against Germany and were in the Army of Occupation."

"I won't let any of your prospects know. I won't mention the war to them. And if I call on the Pfeiffers I'll keep it a dark secret."

"Be sure you do," said George. "Fortunately for me," he went on, "I was not in the Army, so I have one less thing to conceal than you have—which makes it just that much easier. It is always hard not to make a slip when you have so many things to cover up."

"And what are all those things you have to cover up?"

"Well," said George, "the most difficult job is concealing the fact that the Earthworm tractor is made in America."

"Why would you want to do that?"

"To keep them from being prejudiced against it. If I told them that the Earthworm was made in America their nationalistic feelings would at once be aroused. They would look with disfavor on a machine manufactured by their former enemies."

"So what have you been telling them?"

"I have stated that the machine is made in Germany."

"How could you get by with that? All they have to do is look at the nameplate and they can see it is made in Earthworm City, Illinois. Besides, they would know that the Earthworm is an American invention. We had Earthworm tractors over here to pull the artillery in the Army of Occupation. Half the population must remember seeing them. And they couldn't help but recognize them, as there is nothing similar in Germany."

"I thought of all that," said George, "and I have arranged things very cleverly. I have replaced the original nameplates with new ones bearing the word *Regenwurm*, which is German for Earthworm. I admit to everyone that the first Earthworm tractor was an American invention—just like the telephone and the electric light. I also admit that I am an American and that I used to be in the tractor business in America. But I tell them that the machine I am selling is a German improvement, manufactured at Essen by a new company, and so far ahead of the American machine that there is really no comparison between the two."

"And how does this little fairy tale work?"

"Oh, it works fine. It gives me a swell chance to play on their patriotism. I tell them that when I learned that German science, German manufacturing skill and German enterprise had produced such a superior machine, I became completely discouraged with the inferior American article and came over here to sell something I could take real pride in."

"You explain all this to them in German?"

"No, I don't speak German. Some of the prospects talk English, and when they don't I hire an interpreter."

"Well, well, George," I said, "you have certainly gone into this in a very elaborate and interesting way. But it does seem to me that this business of pretending that the machine is German is carrying things a little too far. Somebody will find out that there is no Earthworm tractor factory at Essen and then you'll be in wrong worse than ever."

"Of course we are taking a chance," said George, "but we have to do it on account of the intense anti-American feeling."

"Are you sure," I asked, "that this feeling is as strong as you think it is? Now, when I was here in Coblenz right after the war with the Pfeiffer family we got along fine. The Pfeiffers didn't hate Americans."

"Don't be too sure of that," said Gadget. "They were forced to take you in. They couldn't help themselves. And they probably decided to act sensible and make the best of a bad business and pretend they liked it."

"That might have been so," I admitted.

"Anyway," and George, "there can be no doubt about the popular feeling at present. Everywhere I go they look at me with suspicion. They overcharge me at hotels. They are rude to me on trains, and they try to shortchange me everywhere. You will soon find out that they do the same to you."

"As a matter of fact," I said, "there was a big fat German that was very nasty when we shoved into his railroad compartment on the train down at Frankfurt, and I have a feeling—although I am not sure—that we got gouged on the wagon restaurant."

"You see," said Mr. McGinnis, "as soon as they find out you're an American, they begin to act dirty. And if you're trying to sell them something, they are impossible. I have spent nearly all my time the last two weeks trying to interest a contracting firm here in buying a tractor. They need it to haul dirt for the approaches of a big bridge they are building across the Moselle at Oberzeller, about twenty kilometers from here. But I haven't accomplished anything."

"How have you gone after them?"

"I called six times on the head guy here in Coblenz. He wouldn't even see me. I called five times on the engineer in charge of the work at the bridge. He let me in the last time. But when I told him I had a tractor here and would like to give him a demonstration, he just said he wasn't interested and had me shown out. He acted very suspicious and hostile. They all do, when they find out I'm an American. I'm completely discouraged."

"No use being discouraged," I said.

"But what can we do?"

"I would suggest," I said, "that we all go down and call on this engineer. We'll let Gadget—unhampered by fool interpreters—work on him with all her skillful salesmanship and all her fluency in the handling of the German language. And when she gets through with the guy, I have a feeling that he will be as docile as a lamb."

"I have no such feeling myself," said George, "but I suppose we might as well try. We can go down to Oberzeller on the nine o'clock train tomorrow morning."

"Very good," I said. "What is the name of this contracting firm?"

"The Rheinländische Hochbaukonstructionsgesellschaft."

"Marvelous!" I said. "A company with a name like that ought to buy more than one tractor. They ought to buy three or four. I feel much encouraged."

And with that, our conference ended. This afternoon Gadget and I have been doing a little sightseeing. Tonight we are going to call on my old friends the Pfeiffers. Tomorrow morning we tackle the engineer of the Rheinländische Hochbaukonstructionsgesellschaft. And tomorrow evening I hope to write you of another success to be added to the long list of achievements of

Your hard-working salesman,
ALEXANDER BOTTS.

ALEXANDER BOTTS
EUROPEAN REPRESENTATIVE
FOR THE
EARTHWORM TRACTOR

RHEIN HOTEL, COBLENZ, GERMANY.
THURSDAY EVENING, JUNE 28, 1928.

MR. GILBERT HENDERSON,
EARTHWORM TRACTOR COMPANY,
EARTHWORM CITY, ILLINOIS.

DEAR HENDERSON: Twenty-four hours have passed since my last letter, and I regret to state that we are no nearer making a sale than we were yesterday. Indeed, it may be that the situation is a little worse, owing to the fact that I have made an unfortunate mistake, which seems to be getting us in wrong with potential tractor buyers.

I always try to be perfectly honest and candid in my reports to the home office. When I am going good I always tell you about it. And whenever I make a blunder—which occasionally happens to the best of us—I never attempt to conceal it. Consequently I will tell you all about the events of last night, and I will go into considerable detail so you can see that, although I made a slight error in judgment, I was not very much to blame. In fact, I believe that anyone in my place would have acted exactly as I did.

Directly after supper last night Gadget and I sallied forth to call on the Pfeiffer family. As we walked along, Gadget asked me to be discreet.

"If I were you," she said, "I wouldn't tell these people that you are in the tractor business."

"Why not?" I asked.

"Because they may gossip around with their friends, and before long it might be reported all over town that a former member of the hated American Army of Occupation has arrived for the purpose of selling tractors."

"And that," I said, "would be bad for business. Very good. I will say nothing about it."

We found the house without any trouble—an old stone building on the Goebenplatz right opposite the famous house where Herr Karl Baedeker published his first guidebook over a hundred years ago. As I knocked at the door, I could feel my heart beating fast with excitement. I really had been very fond of these people. It suddenly occurred to me that nine long years had passed. During this time I had had no news of them. What if they had moved away? What if they had died? What if they had come to hate Americans so much that they would no longer regard me as a friend?

The door opened, and there stood old Fran Pfeiffer herself—a little bit fatter than she used to be but otherwise looking exactly the same. At first she did not recognize me. Then she let out one of the gladdest and happiest little yelps I have ever heard. "*Alexander*," she said. "*Der amerikanische Soldat!*" She then grabbed me and kissed me all over both sides of the face. So I gave her a good, long, healthy hug and kissed her a couple of dozen times. It certainly made me feel good. I introduced Gadget, and there was more embracing. Then the good frau dragged us both into the sitting room where Herr Pfeiffer was comfortably installed in his easy chair reading a newspaper and sipping a large mug of beer.

"*Eine grosser Überraschung!*" announced Frau Pfeiffer. "*Unser guter Alexander!*"

Herr Pfeiffer at once sprang to his feet, his whiskers dripping with froth, and his ruddy face lit up with a smile of joy. "*Willkommen! Willkommen! Willkommen!*" he said, and hugged me and kissed me with great gusto. After I had introduced Gadget and was wiping off some of the foam, he gave her a few smacks.

"She then grabbed me and kissed me all over both sides of the face."

The door opened and in came young Wilhelm Pfeiffer, the son of the family. By this time Gadget and I had begun to get into the spirit of the thing, so we grabbed him and saluted him in the approved manner before he even had a chance to realize who we were.

Frau Pfeiffer made several rapid trips to the kitchen and dragged in a lot of bread and *Mettwurst, Leberwurst, Blutwurst. gesalzter Schweinskopf, Gänseschwarzsauer* and plenty of beer for everybody. We then sat down and had a good long visit. And never have I had a pleasanter evening.

As you know, I am not what might be called a drinking man. In general I completely avoid all alcoholic beverages, as I do not wish to dull the keenness of my mind and thus lower the efficiency of my work as a salesman. But in this case I couldn't very well refuse the hospitality of these excellent people. Consequently I partook freely of the food and also of the drink.

And the more I ate, and the more I drank, the more genial I began to feel. The other members of our little group seemed to feel every bit as friendly and genial as I did myself, and before long the party became suffused with a warm glow of complete happiness and good fellowship.

I even began to remember the small amount of clumsy German which I had picked up in the Army of Occupation. I was surprised and delighted to discover how much I knew. But of course I was far inferior to Gadget, so I let her do most of the talking to the old folks, who understood no

English, while I chatted pleasantly with young Wilhelm, who, it appeared, had recently graduated from a university in a town called Bonn, a short distance down the river. I had remembered Wilhelm as an awkward lad about fourteen years of age, who spoke no English at all. He is now a fine-looking young man of twenty-three, with a splendid knowledge of English. We had a long and interesting conversation.

First we discussed the good old days. I recalled how his mother had sewed on buttons and mended clothes for me, and even done most of my washing, and generally looked after me as if I had been one of the family. Wilhelm recalled how I had swiped coffee and sugar and soap from the battery kitchen, and kept them supplied with these necessary articles at a time when there was a great scarcity of them. And he said he would always remember what a thrill he got out of a toy airplane and some toy boats that I made for him.

I said that I was glad to see them all looking so well and apparently so prosperous. He thanked me and said that everything was going very well. His father had saved up a certain amount of money, had retired from business and was living a pleasant and leisurely life exactly to his liking. He—Wilhelm—after graduating from the university had got what he considered a very good position on a newspaper called the *Coblenzer Freie Presse*. Apparently he has done very well in this journalistic work, and his prospects for further advancement are very good.

After telling me all about himself and his parents, Wilhelm wanted to know what I had been doing. So I opened up and gave him a long and glowing account of the great Earthworm tractor and my success as a salesman. Wilhelm remembered the Earthworms which we had used to pull the guns of our battery, and he was naturally very much interested to learn that I was now selling them in Europe. I told him of my adventures in France, and Italy and Russia.

At length Wilhelm had to excuse himself. It appears that the *Coblenzer Freie Presse* is a morning paper, and he had to get down to the editorial office to finish up a certain amount of work before the first edition went to press. Gadget and I stayed around for an hour or so longer. Herr Pfeiffer got out his old accordion and delighted us by playing a lot of American wartime songs which he had learned from the boys in 1919, and had remembered all these years. He played "Over There," "The Long, Long Trail," "The Caissons" and a lot of others. He finished up in a blaze of glory with "Mademoiselle from Armentières," and actually sang us in English a couple of verses which he had committed to memory without

"Herr Pfeiffer got out his old accordion and delighted us by playing a lot of American wartime songs."

having the faintest idea what they meant. These verses startled Gadget, but she was polite enough not to show it.

When, at eleven o'clock, we finally took our departure, we all agreed that we had passed one of the most delightful evenings of all our lives. The Pfeiffers begged us to come in as often as we could during our stay in Coblenz, and said they hoped that in the future we would write to them often so they could know how we were getting along. As Gadget and I strolled home through the streets—which at that hour in Coblenz are dark and almost deserted—my heart was full of contentment and satisfaction.

"Those Pfeiffers," I said, "are regular people sure enough."

"They certainly are," agreed Gadget. "And they certainly seem to be very fond of you."

"They ought to be," I said. "I am very fond of them. And I hope you are going to be fond of them too."

"I am already," said Gadget. "There was one thing, though, that worried me a little."

"What was that?"

"I heard you telling young Wilhelm Pfeiffer that you were a salesman for the Earthworm Tractor Company. I thought we had decided to keep that dark."

"I know," I said, "but he asked me what I was doing, and I couldn't very well lie to an old friend. It won't make any difference now, though. He won't be apt to spread it around where any prospective purchasers might get hold of it."

"Well, I hope not," said Gadget.

Little did we know that at that very moment Wilhelm Pfeiffer was writing up an article about me for the paper.

This morning we took a nine o'clock train with George McGinnis to Oberzeller where they are starting to build the big bridges. After a long delay we finally got into the office of the engineer in charge. This engineer was a severe-looking, businesslike young man who seemed to be considerably annoyed at our presence. He didn't look to me like an easy prospect. Gadget, in her most pleasing and attractive manner, explained in German who we were and what we wanted.

As soon as he heard her mention the name Alexander Botts, he picked up a newspaper from his desk and showed Gadget an article therein. After Gadget had read it he asked her a number of questions. She replied, and they had rather a lengthy discussion, while George and I stood helplessly around, not understanding what was being said or having the faintest idea what it was all about.

We could gather, however, from the cold and unfriendly expression on the face of the engineer that Gadget was not making much of an impression on him. Finally Gadget turned to us and said that there was nothing more we could do. We all filed out of the office and walked back to the railway station. Here Gadget told us about her interview.

"After Gadget had read it he asked her a number of questions."

"In the first place, I want to say that I am disgusted with you, Alexander. You should never have told that Wilhelm Pfeiffer that you are a salesman for the Earthworm tractor."

"Has Alexander been telling people that he is an Earthworm tractor man?" asked George.

"Worse than that," said Gadget. "He has told a reporter, and the reporter has put it all in the *Coblenzer Freie Presse*."

"Was it a long article?" I asked.

"Not very," said Gadget, "but it was plenty. It spilled all the beans—gave your name, said you had fought against Germany in the war, had been a member of the Army of Occupation, and were now attempting to sell the American Earthworm tractor in Germany. When that engineer showed it to me it put me in a very awkward position. He asked me if you were the man mentioned in the paper. As I had already introduced you as the salesman for the German *Regenwurm* tractor I had to tell him that you were an entirely different person but with the same name."

"What did he say?" I asked.

"He said it seemed highly improbable to him that two Americans with the same name should arrive in Coblenz simultaneously, each trying to sell tractors."

"And what did you say to that?" I asked.

"I just repeated what I had told him before—that there were two of you. But I am such a poor liar that I don't think I made it stick. He acted very suspicious and somewhat puzzled. Apparently he thought I was trying to put over some dirty work on him. He couldn't make up his mind what it was exactly, but our interview was ruined just the same. There was a total lack of that mutual trust and friendliness which is always necessary in a successful selling talk. All I could do was give him an unconvincing line about how good this *Regenwurm* tractor is. But he was so suspicious that it made no impression on him at all."

"What did he say?" I asked.

"He said that he had already told Mr. George McGinnis that he was not interested in our machine, that he had not changed his mind, that he did not want to see a demonstration, and would be obliged to us if we would refrain from wasting his time further. He seemed perfectly hopeless, so I wished him good morning and left."

"A fine guy you are," said George, looking at me. "You come up here to help me, and you mess things up worse than ever. We were almost licked before you put your foot into this business. Now we are completely licked."

"No," I said. "I refuse to surrender. I admit I made a mistake in telling that reporter all about myself. I apologize most humbly. But I will do more than apologize. We will return to Coblenz and I will meditate upon the situation and see if I can't figure out some scheme to attack these people in a new way. I have not given up hope."

At these brave remarks Mr. McGinnis laughed rather derisively, and even Gadget looked doubtful. During most of the train ride back to Coblenz I applied my mind to the problem before us, but I could think of no solution. Since arriving at Coblenz I have written this report and taken a long walk on the riverbank. But I still have no promising plan of action. However, I will get a good rest tonight, and tomorrow I may think of something. I will let you know in case there are any interesting future developments.

Very truly,
ALEXANDER BOTTS.

ALEXANDER BOTTS
EUROPEAN REPRESENTATIVE
FOR THE
EARTHWORM TRACTOR

RHEIN HOTEL, COBLENZ, GERMANY.
FRIDAY, JUNE 29, 1928.

MR. GILBERT HENDERSON,
EARTHWORM TRACTOR COMPANY,
EARTHWORM CITY, ILLINOIS.

DEAR HENDERSON: We have had much excitement today. While Gadget and Mr. McGinnis and I were having breakfast this morning one of the bellhops came in and said that Mr. Wilhelm Pfeiffer wished to see me.

"Very good," I said. "Show him up to my room, and I will be there at once."

"I suppose," said Gadget, "he wants another interview for the paper. If I were you, I'd throw him out the window into the river."

"He does deserve a good bawling out," I admitted.

But when I went up to the room, Wilhelm greeted me so cordially that I didn't have the heart to get disagreeable with him. Besides, he had another man with him—a pleasant-looking middle-aged gentleman whom he introduced as an old friend of the family, and a former captain in the German field artillery.

"The captain," said Wilhelm, "saw a little notice about you which I put in the paper, and he was anxious to meet you."

"I am highly honored," I said.

"Mr. Pfeiffer tells me," said the captain, "that you were in the American Army during the war." He spoke very good English.

"I was a private," I replied, "in the artillery. A very humble position," I added, "as compared to your own. You were a captain, I understood?"

"We all had our parts," he replied, "and no one in the war had a more honorable part than the simple soldiers who bore the brunt of most of the hard work."

"Thank you," I said.

"To what organization did you belong?" he asked.

"The Thirteenth Field Artillery of the Fourth Division."

"Let me see. Let me see," said the captain. "Were you near the village of Septsarges about a month before the Armistice?"

"I should hope to kiss a pig," I said. "We were just outside of that little town for over a week. But how did you know?"

"I know," said the captain, "because the battery which I commanded was on your right flank directly across the Meuse River."

"Wait a minute," I said. "Let me get this straight. You say you were in command of the boys that were operating those guns across the river on our right?"

"I certainly was."

"The first week in October, 1918?"

"That was the time. Perhaps you remember that we used to fire at you every once in a while?"

"Do I remember!" I said. "I should say I do. You people had us worried sick. Of all the artillery that we ever ran up against during the whole war, that bunch of shooting fools across the Meuse from Septsarges was the quickest-firing, the most skillful and the most annoying. How you boys used to send your shells screeching down that little valley! And how we used to curse you out!"

"Of course," said the captain modestly, "we always tried to do the best we could. But I honestly believe your Fourth Division artillery was better than anything we had. The way you people used to drop shells all around our battery position was really amazing. I suppose you adjusted your fire by airplane observation?"

"Not at Septsarges," I said. "We were firing entirely by the map."

"Marvelous!" said the captain. "Marvelous!"

"Listen," I said. "I want to ask you a question."

"Certainly."

"Did you ever do any sniping at individual men?"

"Yes. We had direct observation down your little valley from an observation post on one of the hills."

"Aha!" I said. "Then you were the dirty bums that pretty near bumped me off."

"Bumped you off?"

"Yes. I was walking down the road to the east of Septsarges one afternoon when everything had been so quiet that I thought you bozos must have gone to sleep. And then all of a sudden there came over one of those nasty little 77s—so fast I didn't have time to duck, and so close it pretty near took off my tin hat as it went by. It hit right beside the road and blew dirt and sod all over me. The fragments missed me, but I was practically scared to death. For over a week I was a nervous wreck."

"We had 77s in our battery," said the captain. "It is very possible that we fired that very shot."

"You had the devil's own accuracy," I said.

"No," said the captain. "That shot was a failure. We missed you."

"I insist," I replied, "that it was a wonderful piece of work. You were firing at a range of four or five kilometers, by indirect laying, and at a moving target. And I swear that shell went by within three feet of my head. It was magnificent."

"The most magnificent shot of all my experience," said the captain, "was one that you people sent over at exactly four o'clock on the afternoon of October 6, 1918. It was one of those 155 howitzer shells, and it came straight down on top of our kitchen, and burst in the exact mathematical center of a big kettle of soup. The entire kitchen was demolished. All our supplies were destroyed and we didn't have anything to eat for two days. I wonder if it was your battery that fired that shell."

"Very likely," I said. "We had 155 howitzers."

"This is most interesting," said the captain. "Permit me to congratulate you on having been a member of a battery which could achieve such admirable results."

With these words he rose to his feet and grasped me cordially by the hand.

"Sir," I said, "I am sure that the best work our battery ever did—and we certainly worked those old guns as well as we possibly could—was mere amateur bungling as compared to the finished and artistic work of your excellent German battery. I consider it a great honor to shake hands with you. And you don't know what a kick I get out of meeting the man whose remarkable gunnery almost knocked my head off ten years ago."

"The honor and the pleasure," said the captain, "are all mine. How wonderful it is that after all these years I should come face to face with a member of that astounding American battery which could hit the exact center of a kettle of soup at a range of five kilometers."

"This is indeed an auspicious occasion," I remarked, "and it seems fitting that we should celebrate it in appropriate fashion. I trust you will permit me to offer you some light refreshments."

"The idea," said the captain, "is excellent."

"I agree," said Wilhelm Pfeiffer.

I immediately ordered up a reasonable amount of the best Rhine wine and a certain amount of food to go with it, and the rest of the morning was devoted to one of the pleasantest and most interesting little parties I have attended in a long time. Wilhelm, who had been too young to fight in the war, sat around and listened while the captain and I swapped lies and discussed all phases of the great war. At length I happened to mention the fact that our battery had used Earthworm tractors to pull the guns.

"Yes," said the captain. "I remember seeing those tractors in your Army of Occupation up here. And I was present at the Third Army carnival in the spring of 1919, when one of your tractor operators performed the amazing feat of driving his machine straight up the ramparts of Ehrenbreitstein—which have a slope considerably steeper than forty-five degrees. I also remember watching these Earthworms at work around your camps, and I was much impressed at the work they do. They are wonderful. And that reminds me that I have been so interested in discussing the war that I almost forgot the real object of my visit to you. Wilhelm tells me that you are a representative of the Earthworm Tractor Company."

"That is correct," I said.

"Well, I am a contractor and I am interested in buying a few of your machines."

"The captain," said Wilhelm, "is the president of the Rheinländische Hochbaukonstructionsgesellschaft, which is one of the most important firms in the country."

"How interesting," I said.

"We have been thinking for some time of buying a few tractors," said the captain. "The engineer in charge of a bridge we are building down at Oberzeller has been approached by a rather suspicious-looking character who wants to sell us a German imitation of your Earthworm tractor. But of course we don't want any inferior, untested product. What we want is the original Earthworm, built by the same company which produced those excellent machines used by your army."

"That is exactly what I can sell you," I said. "It is the same tractor, although of course it has a certain number of improvements. If you would like to see it in action I can give you a demonstration at any time."

"That will be unnecessary," said the captain. "I will take your word that this machine is as good as the one used in your army. You are a friend of Wilhelm's and an ex-member of a splendid regiment, so I know I can trust you."

At once I got out my literature and specifications on the different sizes of Earthworms, and within half an hour I had completed the sale of four of George's five tractors.

After signing the orders and giving me a check for the down payment, the captain wished me a cordial good morning. Wilhelm invited the captain and Gadget and me to supper at their house this evening, and I accepted with pleasure. The captain and Wilhelm then took their departures.

As soon as they had left, George McGinnis and Gadget came into the room fairly panting with excitement and anxiety.

"The hotel clerk," said Gadget, "told us that the man who has just been visiting you is the high mucky-muck of the Rheinländische Hochbaukonstructionsgesellschaft. What have you been doing with him? I hope that you two haven't been discussing the war."

"And I hope," said Mr. McGinnis, "you didn't let him know that our tractors are really made in America."

"As a matter of fact," I said, "I didn't know who he was until just before he left."

"You didn't!" said Gadget.

"No. All I knew was that he was a friend of Wilhelm Pfeiffer's. So I talked a lot about the war and the army, and I told him that I was a salesman for genuine made-in-America Earthworm tractors."

"Good Lord!" said Mr. McGinnis. "You actually told him all that?"

"Yes," I said, "and then I sold him four machines. Here are the orders, and the check. You will have to take off the *Regenwurm* nameplates and put the original ones back on, because this lad wants real Earthworms. And now that we have helped you out this much. I feel that Gadget and I had better be planning to get back to our regular work in France and Italy. We can't leave tonight—we have a date with the Pfeiffers—but we'll pull out first thing in the morning."

"You sold four tractors!" said George incredulously, and then lapsed into a thoughtful silence. Finally he started for the door. "There's a store here that handles books in English," he said. "I'm going to buy everything they have about the war, and read up until I can talk about it intelligently and at great length. I still have to sell that fifth tractor."

He walked out. And that is all at present from

Yours truly,
ALEXANDER BOTTS.

DEVIL'S GULCH

ILLUSTRATED BY TONY SARG
OPENING ILLUSTRATION BY NICK HARRIS

ALEXANDER BOTTS
EUROPEAN REPRESENTATIVE
FOR THE
EARTHWORM TRACTOR

AT THE FARM OF MONSIEUR PIERRE GROGNARD,
ST. JACQUES-EN-CHAMPAGNE,
MARNE, FRANCE.
TUESDAY, JULY 10, 1928.

MR. GILBERT HENDERSON,
EARTHWORM TRACTOR COMPANY,
EARTHWORM CITY, ILLINOIS.

DEAR HENDERSON: This is to let you know that everything is going swell. You had better forget your absurd letter of recent date, in which you suggested that Europe is such a poor market for tractors that I might as well give it up as a bad job and plan to return to America. As I am always respectful to my business superiors, I will merely state that you are all wrong; I will refrain from explaining exactly how idiotic your ideas really are.

In two or three weeks, when the grain gets ripe, I expect to be selling tractors and combined harvesters by the dozen in the rich wheat-growing region along the Marne. I have already had a demonstration harvester and two tractors shipped up from Marseillas to Château-Thierry. While waiting for the harvest season, I am working on another big idea which came to me only this morning. The circumstances were as follows:

After leaving Germany, I decided to take Gadget to visit the battlefields. Every old soldier naturally likes to look over the terrain on which he fought and did kitchen police for his country, and he likes to take his wife along so he can show her where he performed his most heroic exploits. So I hired a little French touring car, and for several days Gadget and I have been rattling around through the devastated areas. We have seen St.-Mihiel, Verdun and the region around Montfaucon. Last night we stayed at Vienne-le-Château, in the Argonne Forest.

This morning we started for Rheims. But before we had gone more than fifteen or twenty kilometers our radiator began to boil. We stopped beside the road. We investigated. The fan belt was broken. I tried to patch it, but it was too far gone. We were far from any town large enough to have a garage, but a short distance down the road was a stone farmhouse with a number of stone barns grouped around it. We knocked at the door of the house, and were courteously received by an elderly gentleman with large mustaches. He was very helpful, and dragged out various scraps of old leather in the hope that we could make a fan belt out of them. But none of them would work. He told us that there was no place in the vicinity where we could hope to buy a belt, but he said that his son was about to leave on his motorcycle for Rheims, and that he would be glad to buy us one there. The son expected to return in three or four hours.

As this was the best we could do, we accepted his offer. The son departed in a cloud of dust, and the hospitable old gentleman then undertook to while away the hours of waiting by telling us all his troubles. Gadget, being such an excellent French scholar, could, of course, follow his remarks with ease. I could understand most of the gestures and guess at part of the words, and Gadget translated the rest into English as we went along.

Our friend said his name was Pierre Grognard. He had inherited the farm many years ago. "It is not a very good farm," he said, "as you can see for yourselves. It was bad enough before the war, but now it is impossible."

We looked around. The landscape was indeed a bit sad. The land was rough and rocky, and the soil was light-colored and rather thin. There were fields of grass and wheat and cabbages, but they were growing in a feeble, half-hearted manner. Here and there in the distance were patches of scraggly looking trees, and far away to the east we could just make out the wooded heights of the Argonne Forest.

"This region," explained Monsieur Grognard, "is the eastern part of the old province of Champagne. It has always been considered fairly worthless. They even call it *Champagne Pouilleuse* to distinguish it from the rich and fertile port around Rheims."

"And what," I asked, "does *pouilleuse* mean?"

"It means 'lousy,'" said Gadget.

Once more I looked at the dreary landscape. "It is," I admitted.

"Since the war," continued Monsieur Grognard, "it is even more lousy than ever. There was heavy fighting all over this area. I was forced to move to Southern France for the period of the war. Those depressions all over the fields are old shell holes. They make farming very difficult. And so much of the chalky subsoil has been blown up on top that the fertility of the fields is greatly lowered."

"Your farm buildings look all right, anyway," said Gadget helpfully.

"They are not," he said. "Before the war I had good buildings on this place, but they were all blown to pieces. In order to get government aid in rebuilding, I had to employ a contractor and an architect approved by the government. And these pirates, with their technical arguments, talked me into doing everything their way."

"That was too bad," said Gadget.

"It was. I was no match for them. Look, the contractor gave me blue slate roofs, which he said would look as well as red tile. He lied—as you see. And the architect arranged the stables so that I have to throw the manure out the back instead of in front along the road where I could get at it. Yes, everything is wrong, but my chief complaint is that they have not finished clearing up my farm. Come with me, and I will show you my best field, which is still of no use to me on account of the slowness of the imbecile officials who have charge of the rehabilitation of the devastated areas."

He led us around the house to a sort of stone terrace at the rear, from which we looked down into a deep and narrow little valley. The sides were steep and wooded, and the bottom was a flat field of fifteen or twenty acres covered with luxuriant grass.

"What a pretty little valley," said Gadget, "and what a fine field for growing crops. It looks as if the soil down there must be very rich."

"It is," said Monsieur Grognard. "Before the war it was the most fertile piece of ground in this whole region. It was capable of producing twice as much wheat as all the rest of my farm put together. But since the war it has been of absolutely no use to me whatsoever."

"What's the matter with it?" asked Gadget. "It looks all right to me."

"In a sense it is all right. It is still just as fertile as ever. The soil is so deep that the bursting shells did not reach the subsoil and blow it up to the surface."

"Then why can't you grow crops there just the same as ever?"

"Because," said Monsieur Grognard, "I don't dare plow that ground. It is all full of unexploded shells."

"How annoying."

"Annoying! It is insupportable. And it is unjust. The government is supposed to use all the money from the German reparations in rehabilitating the devastated areas. And as the reparation payments have been pitifully small, the government has set aside vast additional sums for this work. They have plenty of money. They have plenty of workmen. They have cleaned up other farms. Why don't they clean up mine?"

"They did clean up some of your fields, didn't they," Gadget asked, "and they put up new buildings?"

"Yes," admitted Monsieur Grognard, "but what good does that do me? The fairest, the most beautiful and the most productive part of my farm they have left untouched."

"Isn't there anything you can do about it?"

"I fear not. I am greatly discouraged. I have pleaded with them. I have lodged protests with the higher officials in Paris. But still they do nothing but make excuses."

"I should think," said Gadget, "that by this time they ought to have all the reconstruction work finished. It is almost ten years since the war."

"Yes. But the government is slow. It took them a long time to get started, and the work has proceeded at a very gentle speed. There is still a great deal to be done, both here and in the other battle areas."

It was at this point that my natural selling instincts commenced to wake up. Before this visit to the devastated regions I had supposed that all the work of cleaning them up had been finished. But it now appeared that such was not the case. And if there was still a lot of work to be done, there might be a need for a certain amount of machinery to expedite it.

"They have their excuses," Monsieur Grognard continued. "They say the work in this valley would be so dangerous that they may never attempt it at all. Where there are only a few shells they get along fairly well. They send out workmen who go over every inch of the fields, Wherever a German shell has entered the ground and failed to explode, they can usually find it by the small hole—like a rabbit burrow—which it made. After these holes are located it is a simple matter to dig up the shells, place them in piles, and explode them with charges of dynamite."

"Do they ever have any accidents?" asked Gadget.

"Sometimes."

"It seems hardly possible," said Gadget. "If a shell fails to go off after it is fired from a gun, whizzes through the air for several miles and hits the ground with a terrific jolt, I should think it never could go off—especially after lying in the wet ground for ten years."

"It seems incredible, madame, but such things occur. Occasionally those shells will go off at the mere touch of a plow point or a shovel. Only last week one of the government workmen was killed over in the Argonne Forest while he was digging up one of these treacherous projectiles. It is very sad."

"And yet the work goes on?"

"Yes, madame—everywhere except on my farm. They claim I have too many shells. You see, the history of this little valley is most curious."

"How so?"

"Well, early in the war they established a division headquarters down there. The spot was naturally somewhat protected, and they improved it by doing a lot of digging. The generals and other officers had large dugouts along the north side of the valley, and the simple soldiers of the headquarters detachment made themselves smaller ones all over my beautiful field. How disgusting it was! With so much worthless land all around, they nevertheless chose to desecrate my loveliest field in that way. But that was not the worst."

"No?"

"In the course of time the Germans became more active, and this valley was too near the Front for the comfort of the generals. The headquarters was moved back and the place was used for one of the largest ammunition dumps in the entire sector. The son of one of my neighbors was here during the war and he has told me about it. It was terrible. They brought in vast numbers of shells and huge quantities of fuses, signal rockets, machine-gun cartridges, powder cases, hand grenades—everything. And they piled the whole grisly mass in my charming valley. They not only covered the surface of the ground but they filled up all the dugouts. And then the Germans discovered what had been done."

"And what did the Germans do about it?" asked Gadget.

"They gave my unfortunate valley a terrific and long-continued bombardment every day for weeks. The supplies that were piled on top of the ground were utterly destroyed, pulverized, annihilated. The surface of the earth was churned up and blown hither and thither, backward and forward. Some of the ammunition stored in the dugouts was destroyed, but much of it was merely buried, so that there are thousands and thousands of French artillery shells and enormous quantities of miscellaneous explosives incorporated in the soil. As all the landmarks were destroyed and all the entrances to the dugouts completely obliterated, no one knows the exact location of these ghastly deposits. And that is not all."

"I don't see how it could be much worse," said Gadget.

"In addition to the French explosives," continued Monsieur Grognard, "there are probably many thousands of German duds. The bombardment of my little valley took place toward the end of the war, at a time when the German ammunition was of very poor quality. The son of my neighbor who was here tells me that at least a third of the shells which the Germans fired at that time failed to detonate."

"What a frightful mess," said Gadget. "I can't say that I would blame the government workmen for being a bit shy about stirring up any such hornets' nest as that place must be."

"Well," admitted Monsieur Grognard, "I don't want to fool with any such job myself."

"And apparently the government doesn't want to do anything either?"

"They do not. And I fear it is hopeless. I must make up my mind to accept this crushing misfortune. Never again can I grow any crops on that land, which was once so rich and productive. And never again can I be a prosperous farmer, for that field was the cream of my possessions."

"It seems a shame," I said, "to give up so easy."

"But what can one do?" said Monsieur Grognard. "And what is man but a mere insect tossed about on the sea of life by forces which he is powerless to resist?"

"I don't know about you," I said, "but certainly I am no helpless insect tossed about on the sea of life."

"When the fates are against us it is useless to struggle. What frail, impotent creatures we are. Today we are here. Tomorrow—*pouf*—we are gone."

"Don't be a boob," I said. "If you don't like the way things are going you ought to get busy and do something about it. Suppose this is a tough proposition. That is no excuse for laying down and moaning about insects on the sea of life."

Monsieur Grognard looked a little hurt. "I was merely trying to view my troubles from the point of view of a philosopher," he said.

"What you need to do," I remarked, "is to view them from the point of view of a practical man like me." All this conversation, of course, moved rather slowly, as Gadget had to interpret almost all of what Monsieur Grognard and I said to each other. "It just happens," I continued, "that your problem here interests me very much. I am a tractor salesman, and I have just been inspired by a great idea. I am going to bring one of my machines up here and I am going to clean up your field. I am going to invite the government officials to watch me do it. and then I am going to sell them a whole lot of machines to do similar work elsewhere. Isn't it marvelous?"

"It would be," said Monsieur Grognard, "if you could do it."

We all went into the house, and I had a long conference with Gadget. At first she was all against my idea. She was afraid I might get myself blown up. But I finally convinced her that the danger would not be excessive, and she agreed that we might as well go ahead and see what we could do. Monsieur Grognard very kindly invited us to lunch, and I have been spending the early part of the afternoon writing this report. It is now four o'clock, and Grognard *fils* has just returned with the fan belt, which we have installed on the automobile. Gadget and I are leaving at once for Château-Thierry, where we trust the two tractors have already arrived. We plan to drive one of them up here tomorrow, and within two or three days we expect to be all set for one of the most interesting and sensational tractor demonstrations in the entire history of the business. I will keep you informed as to our progress.

Most sincerely,
ALEXANDER BOTTS.

ALEXANDER BOTTS
EUROPEAN REPRESENTATIVE
FOR THE
EARTHWORM TRACTOR

AT THE FARM OF MONSIEUR PIERRE GROGNARD,
ST. JACQUES-EN-CHAMPAGNE,
MARNE, FRANCE.
FRIDAY, JULY 13, 1928.

MR. GILBERT HENDERSON,
EARTHWORM TRACTOR COMPANY,
EARTHWORM CITY, ILLINOIS.

DEAR HENDERSON: Since my last report we have put in several days of intense activity. Late Tuesday afternoon Gadget and I drove our little rented car to Château-Thierry. On Wednesday I drove one of the tractors up here to old man Grognard's farm, while Gadget brought the car. On Thursday—yesterday—we scouted around over the nearby battlefields and picked up a lot of the old elephant iron which was used during the war to make the roofs of dugouts and bombproof shelters. With this iron we built a rough armored box, completely enclosing the driver's seat and the motor and radiator of the tractor. This morning we went to Rheims, where I purchased several cases of champagne, a box of dynamite with caps and fuse, a hundred feet of heavy wire cable, and a couple of big subsoil plows with hooks that go down about a meter—more than three feet—into the ground. When we got back to the farm with this apparatus we found that Monsieur Grognard—following our instructions—had invited various contractors and officials to a grand demonstration which we are planning to put on tomorrow.

I had intended to have the big show today. But Monsieur Grognard would not hear of it. Apparently he still thinks of himself as an insect floating on the sea of life, and as this is Friday the thirteenth, he is afraid to tempt the fates by starting any enterprise on such an inauspicious date. We therefore decided to humor him. Furthermore, tomorrow is the glorious Fourteenth, the day that Napoleon captured the Bastille or something, and the chief national holiday of France. It will be most fitting for us to give our splendid performance on this notable day.

And in my next letter I hope to write you of the success of our demonstration and of the purchase by the French Government of an adequate fleet of Earthworm tractors to be used in the great humanitarian work of cleaning up the pernicious remains of war and restoring to the sweet uses of peace even the more lousy portions of *Champagne Pouilleuse*.

Yours hopefully,
ALEXANDER BOTTS.

ALEXANDER BOTTS
EUROPEAN REPRESENTATIVE
FOR THE
EARTHWORM TRACTOR

AT THE FARM OF MONSIEUR PIERRE GROGNARD,
ST. JACQUES-EN-CHAMPAGNE,
MARNE, FRANCE.
SATURDAY, JULY 14, 1928.

MR. GILBERT HENDERSON,
EARTHWORM TRACTOR COMPANY,
EARTHWORM CITY, ILLINOIS.

DEAR HENDERSON: We haven't sold any tractors yet, but we still have high hopes. The various contractors and government officials arrived about eleven o'clock this morning, and Gadget made them a long and eloquent address. After flattering them by graceful references to the glorious holiday which we were celebrating, and by complimentary remarks about the French nation, she gave them a history of the little valley which we were going to clean up, and explained how amazingly full of unexploded shells it was. She then told them exactly what we purposed to do.

"Our plan," she said, "is simple but effective. We are going to plow the entire surface of this field with a large subsoil plow which will bring to the surface all shells which are buried to a depth of three feet or less. Shells deeper than that do not matter; they are entirely harmless because no ordinary plowing or cultivating would ever touch them. We would not dare to plow this field in the usual way, because occasionally these unexploded shells go off at a mere touch and plow, plowman and horses would be blown to atoms. It is our plan to pull our plow at the end of a thirty-meter cable. If we strike a shell and it explodes, the plow may perhaps be damaged, but the tractor will be at a safe distance. And the tractor operator will be protected from flying fragments by the armor plate which we have put on the machine. The plow is of the wheel type, which requires no plowman to steer. You will note that a tractor is absolutely essential. You must have a heavy plow that goes deep. Such a plow would require at least a dozen horses—so many that it would be almost impossible to protect them from the deadly, whizzing fragments. With our tractor this matter of protection is very simple. We will now start the demonstration."

Gadget led the officials around to the little terrace overlooking the valley and began to put them in a good humor by serving them the refreshments we had brought from Rheims. In the meantime I rushed down to the tractor and plow, which I had previously parked at one end of the valley. Monsieur Grognard's son—the one that owns the motorcycle—is a very good mechanic. He had become much interested in the tractor and insisted upon accompanying me. We took our places on the seat inside the armor, and I drove bravely out and straight across the field which was reputed to be so full of deadly explosives. I will cheerfully admit that I was a bit scared, and I was much relieved when we got across to the far side safe and sound.

As we were pulling the plow on the end of a thirty-meter cable, I had to drive the tractor thirty meters into the woods in order to bring the plow to the edge of the field. I then turned the tractor around and young Grognard and I got out and walked back over the furrow. We found that, in addition to several helmets and other souvenirs, we had turned up four small shells and one large one. Gently and with infinite care we gathered up the four little ones and piled them beside the one large one. Then I put a half stick of dynamite right opposite the nose of the big baby, inserted a cap and one end of the fuse and lit the other end. Young Grognard and I retreated to the other side of the tractor, and half a minute later the dynamite and the entire heap of shells went off with a louder noise than anything I have heard since 1918.

We then drove back, making a second furrow, and this time we turned up six shells, which we duly exploded with another half stick of dynamite. After that we continued plowing back and forth, exploding shells after each trip, until three in the afternoon. We were so excited that we forgot all about lunch. As young Grognard was greatly interested in the tractor, I let him do most of the driving. This pleased him and it also gave me an opportunity to take some photographs of the work.

At three o'clock we knocked off and had the spectators come down to look over the tractor and check up on what we had done. We found that in four hours we had covered perhaps an acre, and had plowed out and exploded about a hundred shells. Fortunately, not one of them had gone off when the plow struck it.

The officials and contractors seemed very much pleased and impressed. They were good enough to say that our methods were infinitely faster, safer and more thorough than anything they had ever seen before. But when Gadget started in on a rapid-fire sales talk she got no reaction at all. The contractors, it appeared, were mostly builders, they were not interested in doing any such dangerous work as this, and said it was the business of the government to handle it. The officials were all from the local *arrondissement*; they said they had no authority to make purchases, and suggested that we see the high dignitaries of the department at the Prefecture in Châlons-sur-Marne.

As there seemed to be nothing else we could do, we decided we might as well follow this advice. One of the local boys gave us a letter to a man he knows in the Prefecture. We plan to spend Sunday here, and drive down to Châlons on Monday morning. During our absence young Grognard will keep on with the work, so that it will be in full swing in case we want to bring up any of the big bugs from the main office.

Our selling campaign, although halted for the moment, is not permanently stopped. And before long I hope to write you of our complete success.

As ever,
ALEXANDER BOTTS.

ALEXANDER BOTTS
EUROPEAN REPRESENTATIVE
FOR THE
EARTHWORM TRACTOR

HOTEL DE LA HAUTE-MÈRE-DIEU,
CHÂLONS-SUR-MARNE, FRANCE.
MONDAY NIGHT, JULY 18, 1928.

MR. GILBERT HENDERSON,
EARTHWORM TRACTOR COMPANY,
EARTHWORM CITY, ILLINOIS.

DEAR HENDERSON: Nothing much to report except that we saw the guy at the Prefecture. He said he had no authority to buy anything, and referred us to Monsieur Albert Legendre, in the purchasing bureau of the Ministry of the Liberated Regions at Paris.

We leave for Paris in the morning.

As ever,
ALEXANDER BOTTS.

ALEXANDER BOTTS
EUROPEAN REPRESENTATIVE
FOR THE
EARTHWORM TRACTOR

GRAND HÔTEL ROYALE SPLENDIDE
ET DE L'UNIVERS

Ascenseur
Chauffage Central
Eau Courante—Chaude et Froide

Nettoyage par le Vide
English Spoken
Téléphone

RUE ST. HONORÉ, PARIS.
TUESDAY NIGHT, JULY 17, 1928.

MR. GILBERT HENDERSON,
EARTHWORM TRACTOR COMPANY,
EARTHWORM CITY, ILLINOIS.

DEAR HENDERSON: We drove down from Châlons this morning. This afternoon we saw Monsieur Albert Legendre. He said that jobs like cleaning up unexploded shells were the business of the military authorities, and advised us to consult a man called Capt. Auguste Schmitt, in the Ministry of War. We called at Captain Schmitt's office, and were told he would see us tomorrow. This evening we improved our minds by attending the opera. It was not much good.

LATER, WEDNESDAY, JULY 18, 1928.

This morning we saw Capt. Auguste Schmitt. He is not a German. Many of these frogs have Dutch-sounding names.

He was very cordial, and Gadget warmed up to the extent of giving him about a two-hour sales oration, illustrated with the photographs we had taken on Monsieur Grognard's farm. She also pretty near covered him up with advertising leaflets and folders.

He finally said he was much interested, but as long as the work we had in mind concerned agricultural land, it would come under the Ministry of Agriculture as well as of War. He said he would have to see some of the agricultural experts, and he told us to come back tomorrow.

In the afternoon we visited the Eiffel Tower and the Tomb of Napoleon, and saw all the pictures and statues in the Louvre. In the evening we went to the Folies-Bergère. It was better than the opera.

THURSDAY NIGHT, JULY 19, 1928.

This morning, when we called on Captain Schmitt, he appeared surprised to see us. He was also slightly embarrassed and considerably amused.

"What a joke it is," he said. "Your affair had entirely slipped my mind. I have not yet seen the agricultural authorities, but I will surely do so this afternoon. Come back tomorrow."

We withdrew politely. In the afternoon we visited the Bois de Boulogne and the Tomb of the Unknown Soldier. In the evening we went to the Casino. Paris seems to be quite a town, but apparently not much of a place for people like us to do business in.

FRIDAY NIGHT, JULY 20, 1928.

This morning Captain Schmitt said he had seen the agricultural officials, but they had decided that they could do nothing until they conferred with the Ministry of Labor. As it would take some time to arrange this conference, he told us not to come back until Monday. He then ushered us out of his office. I will have to admit that these delays are beginning to get on my nerves. However, if we have to wait around, I suppose we might as well wait in Paris as anywhere else. This afternoon we made a grand tour, visiting the Café du Dôme, the Café de la Rotonde, the Luxembourg Museum, the Père-Lachaise Cemetery, and the Gare de l'Est. Now that we are getting used to this sightseeing, we find that we can do it much more quickly and efficiently, and cover much more ground in a shorter time, than when we first began. This evening we went to the Moulin Rouge.

SATURDAY, JULY 21, 1928.

Today we went to so many places I can't remember any of them. We are certainly getting to be real sightseers.

SUNDAY, JULY 22, 1928.

More sightseeing. I hope Captain Schmitt has something definite to tell us tomorrow. We can't stick around here forever. Paris is a good town, but the wheat up around Château-Thierry is almost ripe, and I ought to be there right now getting ready for our demonstration with the combine harvester.

MONDAY, JULY 23, 1928.

This morning Captain Schmitt was out, but late this afternoon he came back to his office and gave us some very definite news.

"In the first place," he said, "we have decided that we cannot buy any foreign-made goods, as they would tend to take work away from the men in the factories of France. We cannot encourage outside competition with the automotive industry of France."

"But we aren't competing with anybody in France," said Gadget, "because there is no French factory that is making tractors suitable for this work."

"In the second place," continued Captain Schmitt, "we don't want labor-saving machinery. It would take work away from worthy French citizens who need their jobs to support themselves and their families."

"That argument," said Gadget, "is completely phony."

"In the third place," he went on, "we don't need your tractors anyway. The army owns a lot of tanks, and we have decided to use them in this work."

"In other words," said Gadget, "your first two reasons don't amount to anything. It's the third one that counts."

"No doubt you are right. We are adopting your method of cleaning up these dangerous fields. You have our profound thanks. But we don't need your machinery."

Gadget tried to put up an argument, but it was no use.

"Our decision is final," said Captain Schmitt. "There is nothing more to be said. I wish you a most cordial good afternoon."

So that was that. As there doesn't seem to be anything more we can do here, we are leaving in the morning and driving back to the farm of Monsieur Pierre Grognard at St. Jacques-en-Champagne. Much as we hate to do so, I suppose we shall have to take the tractor away and proceed to Château-Thierry in the hope that our next venture may turn out more fortunately than this one. Needless to say, we are very low in our minds.

Before closing this letter I wish to mention a rather curious thing which we ran into this evening, and which has puzzled us quite a bit. After supper an American gentleman who is staying here at the hotel started telling us of a highly interesting motorbus tour of the battlefields which he had taken on Saturday.

"You ought to take this bus ride," he said, "and you ought to take it right away, and be sure to go on the same line that I took. They are making a special side trip to a place where some very dangerous work in cleaning up the battlefields is being done. Here is some literature they gave me that tells all about it."

He handed me a small leaflet—printed in English for the convenience of American tourists—which read as follows:

COMPAGNIE GÉNÉRALE DES GRANDE AUTOBUS ET DES SUPER-SUPERBE TOURS

EXTRA SPECIAL! FOR A SHORT TIME ONLY

During the next few weeks, patrons of the Super-Superbe Tour of the battlefields will be taken to visit the famous *Gouffre du Diable*—Devil's Gulch. During the war this little valley was the site of one of the largest ammunition dumps on the entire Western Front. Enormous quantities of ammunition were stored on top of the ground and also in the vast subterranean chambers and galleries. In 1918 the place was subjected to incessant bombardment by the Germans. Part of the ammunition was destroyed, but great quantities were merely burned. And today the soil in the bottom of the *Gouffre du Diable* is literally impregnated with millions of deadly projectiles, many of them of such a construction that a mere touch is sufficient to explode them with appallingly disastrous results.

It is this valley of death that is now being reclaimed. Patrons of the Super-Superbe Tours are taken to a safe point of vantage from which they can look down into this inferno and see a gigantic armored tractor clanking back and forth, pulling a special steel bombproof plow which digs up all shells and frees the ground so that later it may be used for agricultural purposes. Some of these shells are harmless. But most of them explode at the first touch of the plow.

See this awe-inspiring spectacle! Every few seconds an ear-splitting report, with hundreds of deadly fragments flying through the air! A scene as near like the Great War as can be found anywhere on earth! An exclusive feature offered by no other sightseeing company than

LA COMPAGNIE GÉNÉRALE DES GRANDS AUTOBUS ET DES SUPER-SUPERBE TOURS!

The reading of this pamphlet puzzled me, as I have said, very much. The whole thing sounded very much like a description of the work we were doing in the little valley owned by Monsieur Grognard, except that this was so incredibly more spectacular. I asked the American gentleman exactly where this valley was located, but he had only a vague idea that it was up somewhere the other side of Rheims.

"Did you actually see these shells going off every few seconds as the plow hit them?" I asked.

"Absolutely," he said. "And it was one of the most remarkable sights I have ever seen. It is the sensation of the hour. All the American tourists in Paris have heard about it, and everybody wants to go and see it. The bus company is doing terrific business. You ought to take the trip yourself."

"I haven't time now," I said, "but tomorrow I am driving up in that direction in my own car, and I hope I may be able to learn some more about this curious business."

If I do learn anything about it, I will let you know.

Very truly,
ALEXANDER BOTTS.

ALEXANDER BOTTS
EUROPEAN REPRESENTATIVE
FOR THE
EARTHWORM TRACTOR

AT THE FARM OF MONSIEUR PIERRE GRONARD,
ST. JACQUES-EN-CHAMPAGNE,
MARNE, FRANCE.
TUESDAY, JULY 24, 1928.

MR. GILBERT HENDERSON,
EARTHWORM TRACTOR COMPANY,
EARTHWORM CITY, ILLINOIS.

DEAR HENDERSON: Gadget and I left Paris this morning in our little French car and arrived here early this afternoon. As we stopped in front of the farmhouse old Monsieur Grognard came rushing out to welcome us, his face lit up with a glad, happy smile.

"Welcome back!" he cried. "How glad I am to see you! How has everything been going in Paris?"

"Rotten," said Gadget. "We couldn't seem to make any impression on them at all. We are completely discouraged. How has everything been going up here on the farm?"

"Splendid!" said Monsieur Grognard. "Magnificent! When you see the remarkable impression your tractor is making here, you will not care whether it makes an impression on the people in Paris or not."

At this moment we heard a grinding of brakes behind us and turned around to observe a large motorbus with the sign COMPAGNIE GÉNÉRALE DES GRANDS AUTOBUS ET DES SUPER-SUPERBE TOURS.

"Here is another load," said Monsieur Grognard. "Come with me and I will show you how we handle them."

He took us around the house to a little gate which led to the terrace behind the house. Over this gate was a large sign which said GOUFFRE DU DIABLE—ADMISSION, 10 FRANCS. Monsieur Grognard had me step through the gate while he remained behind and collected ten francs from each of the bus passengers who came in. The English-speaking guide who accompanied these people emitted a short ballyhoo about the remarkable sight we were about to see, and then we all looked down into the valley.

We saw the tractor come crawling out from the woods on one side and drive straight across the field, pulling the plow at the end of the thirty-meter cable. Suddenly there was a puff of smoke from the neighborhood of the plow, and a terrific report which echoed back and forth from the sides of the valley in a truly terrifying manner. A few seconds later there was another puff of smoke and another terrific report, and this continued all the way across the field. There must have been at least a dozen of these explosions. The spectators all around me were impressed, thrilled and tremendously excited; and I will have to admit that Gadget and I got quite a kick out of it ourselves.

After the tractor had made one trip the guide announced that it would have to stop for the purpose of making repairs to the plow, but that the party would remain a few minutes in case any of them wished to have refreshments or buy souvenirs. A good many of them decided that refreshments would be a good thing, and they were promptly served—at a very good price—by Monsieur Grognard and his wife. Meanwhile one of the hired men was doing a brisk business in war souvenirs—shell fragments, German helmets, and such things—at a counter near the gate. When the busload of tourists finally resumed its journey, Monsieur Grognard turned to Gadget and me and smiled broadly.

"How can I ever thank you," he said, "for all you have done for me?"

"I don't understand what we have done for you," said Gadget, "and I don't understand about all these tourists, and all this *Gouffre du Diable* business."

"I will explain," he said. "I am grateful to you people because you have shown me a new philosophy of life. And it works. When you first came I thought of myself as a mere insect on the sea of life. I felt it was useless to

"There must have been at least a dozen of these explosions. The spectators all around me were impressed, thrilled and tremendously excited."

struggle against fate. But you showed me, by the way you started to clear up my little valley, that many a problem which seems hopeless can often be solved. I decided to follow your system, and the very day you left I had a chance to put it into operation."

"What happened?" asked Gadget.

"My son was plowing in the bottom of the valley. A motorbus stopped in the road before my house to repair a flat tire. The passengers got out. They walked around. They stood on my terrace. And as they were looking down into the valley the plow struck a shell which exploded. The passengers were mostly American tourists, who, as you know, are in many ways like children. They love novelty. They love a large noise. They love anything spectacular. They love anything that looks dangerous. They were highly excited and tremendously pleased. In fact, our little explosion was such a success that the conductor of the bus came to me privately to see if I could arrange to have a similar explosion for the amusement of his passengers every day."

"What did you say to that?" asked Gadget.

"In the old days, madame, I would have explained to him that it was impossible—that the explosions were accidental, and that we could not guarantee to have them at any particular time But then I thought of you, and your way of looking at things in a practical manner. I decided that 1 had been too long an insect floating on the sea of life. I decided to act like a superior being—like a lion, an elephant, an American tractor man—in short, like one who controls his environment, rather than letting his environment control him."

"And what did you do?"

"I spoke up in the same masterful manner which I had so much admired in your husband. I said I would furnish all the explosions desired, and would put on a show that would be a veritable sensation. I told him I would give his company the exclusive rights to bring people to the *Gouffre du Diable*—which is a name I made up myself on the spur of the moment. I stipulated that I should be allowed to charge admission and that he should advertise the performance adequately in Paris. He agreed to this, and the arrangement, as you see, has been a tremendous success."

"But what I don't understand," said Gadget, "is how you are able to run into those shells so that they will explode at exactly the right time and place. And I don't see how the plow can stand such punishment."

"Those explosions are not shells at all," said Monsieur Grognard. "They are sticks of dynamite which we plant out there ahead of time and which are set off by electricity. We have one of the hired men concealed in the woods with a magneto which is connected by wires to the dynamite. It works most beautifully, and the customers just eat it up.

"The bus company has chartered extra machines and they are bringing as many as twelve and fifteen busloads of tourists every day. I am making money far more rapidly and easily than I could ever hope to at farming. And it is all due to you two people. You are marvelous. You are magnificent. You are colossal."

"Thank you," said Gadget. "It really does us a lot of good to hear you say that. It restores our self-confidence. Perhaps you won't believe it, but we were so discouraged because we had totally failed to sell any tractors on this particular selling drive that we were actually beginning to think of ourselves as insects floating on the sea of life."

"But you have not totally failed," said Monsieur Grognard, "because I myself wish to buy the tractor which is here. I have a little money of my own, and the bus company is going to advance me the balance of the purchase price. If you would consent to sell it to me. I can pay you the full value in cash at once."

"Nothing would give us greater pleasure," said Gadget.

And I might add that the pleasure is shared by

Your hard-working salesman,
ALEXANDER BOTTS.

HORSE PLAY

ILLUSTRATED BY TONY SARG
OPENING ILLUSTRATION BY NICK HARRIS

Alexander Botts
European Representative
for the
Earthworm Tractor

Hôtel Jean-de-la-Fontaine,
Château-Thierry, Aisne, France.
Saturday, July 28, 1928.

Mr. Gilbert Henderson,
Earthworm Tractor Company,
Earthworm City, Illinois.

DEAR HENDERSON: The last few days have been very busy. We arrived in this magnificent wheat-growing region last Wednesday, and at once started making inquiries to find out what would be the most promising place to put on a harvesting demonstration. Gadget, of course, did most of the talking and inquiring, while I tagged along, congratulating myself on having a wife who is such a fine French scholar.

The first information we received was rather discouraging. Although this region produces large crops of wheat—as all members of the A.E.F. who were here in 1918 will remember—most of the farms are small and the fields are so cut up that they are not very well adapted to large-scale harvesting methods such as we use on the vast plains of Kansas and Nebraska. I began to fear that I had made a mistake in coming to this part of France. But as long as I was here with a perfectly good Earthworm tractor and a full-sized Earthworm combined harvester, I decided to keep going and do the best I could.

We continued our investigations, and finally learned that about ten kilometers from here there is an old castle known as the Château de Moequethon, which has very extensive grounds, including some of the largest wheat fields of the region. We decided this place might offer possibilities.

On Thursday we drove out in the little French touring car which we have rented, and called on the proprietor.

He turned out to be a large blond gentleman by the name of Monsieur Georges Cru. We found out later that he is a well-known magnate in the French moving-picture business.

He didn't know any English, so Gadget talked to him in French. He told her that he had a hundred hectares—about two hundred and fifty acres—of wheat, all in one field, which he wished to start harvesting next week. He was much interested in our account of the Earthworm tractor, and of the Earthworm combined harvester, and said he would be delighted to have us demonstrate it on his farm. He said he would be inclined to buy our harvester if it worked as well as we said it did, and he promised he would invite a number of the more prominent landowners of the neighborhood to be present, so that they also could see the machine in action. He then gave us some rather interesting news.

"This will be in the nature of a competitive demonstration," he said.

"How so?" asked Gadget.

"Several weeks ago I was approached by a man who wished to sell me a combined harvester which is made by a French concern down at Lyons. This man is bringing his machine up here to give an exhibition of harvesting next Monday."

"You surprise me," said Gadget. "I didn't suppose that combined harvesters were made anywhere on earth but in America. But maybe this is not a complete harvester. Are you sure it isn't just a binder?"

"No," said Monsieur Cru. "The man says it cuts and threshes the grain just as you tell me your machine does. This French harvester is an entirely new product. But it is made by an old agricultural-implement concern, so it ought to be all right."

"Well," said Gadget, "it will have to be very good indeed if it is going to compete with the wonderful Earthworm combine."

"You seem to be very confident," said Monsieur Cru.

"We are," said Gadget. "If it is agreeable to you we will bring our machine here tomorrow and get it set up and adjusted. Then on Monday we will start harvesting in the same field as the French machine, and we will let you be the judge as to which is the better apparatus."

Monsieur Cru said that this plan would be entirely satisfactory to him, so we took our departure.

On Friday we went to the freight warehouse and got out the harvester and the tractor. I drove them out to the Château de Moequethon while Gadget followed along in the little car. We spent the afternoon greasing and adjusting the tractor and the harvester, and returned here for the night.

This morning—Saturday—we drove out again to give the machinery a last inspection and make sure everything was all right. When we got to the Château we were much interested to find that the great French harvester had arrived. Gadget and I at once ran over to see it. We found a rather attractive boy about seventeen years old working on it, and we introduced ourselves and said we would like to look it over. The boy told us that his name was André, that he was a sort of assistant mechanic and that he would be delighted to show us anything he could about the machinery. We thanked him, and he then very kindly explained everything to us. And seldom have Gadget and I seen a more pathetic piece of machinery.

Young André was so polite and obliging, and so enthusiastic, that we didn't have the heart to disillusion him, but the truth is that this French harvester seems to be about as far advanced as the Earthworm harvesters of the year 1897. It is designed to be pulled by horses. The threshing-machine part is geared up to one of the rear wheels so that all the power has to come from the motion of the machine as it is dragged across the field. André admits that this makes it very hard to pull. He says they have to use at least thirty horses, and they move very slowly. I doubt if they have enough power to thresh the grain even half as fast as we do with our special motor to run our threshing machinery and our powerful tractor to do the hauling.

As we were finishing our inspection of this quaint piece of machinery we saw approaching a little shrimp of a man whose appearance was most unprepossessing. His complexion looked slightly bilious, he had little weasel eyes and the whole expression of his face seemed to denote craft and insidious cunning.

Little André introduced him to us as Monsieur Jean Jacques Leboutellier, the sales representative in charge of the French harvester machine. As soon as Monsieur Leboutellier learned who we were he became very insulting, accused us of spying upon his machine with the purpose of stealing trade secrets and ordered us to move away at once. Gadget politely told him that he need have no fears, as we could obviously learn nothing of any value from such a poorly made piece of junk. This only served to make him madder than ever, so we smiled pleasantly and walked back to our own machine. As we left we could hear him delivering an unmerciful bawling-out to poor little André for permitting us to look over the harvester.

On the way back to Château-Thierry later in the afternoon, Gadget and I both commented on the fact that in every country we have visited—America, Italy, Russia, Germany and France—most of the people are attractive, decent and reasonable, but that mixed in with these good people there exist in every one of these countries a few individuals of the Leboutellier type.

Tomorrow—Sunday—we are going to devote to rest and relaxation. And on Monday we expect to have the great pleasure of making a complete monkey out of Monsieur Jean Jacques Leboutellier.

Yours, as ever,
ALEXANDER BOTTS.

ALEXANDER BOTTS
EUROPEAN REPRESENTATIVE
FOR THE
EARTHWORM TRACTOR

HÔTEL JEAN-DE-LA-FOUNTAINE
CHÂTEAU-THIERRY, AISNE, FRANCE.
MONDAY EVENING, JULY 30, 1928.

MR. GILBERT HENDERSON,
EARTHWORM TRACTOR COMPANY,
EARTHWORM CITY, ILLINOIS.

DEAR HENDERSON: Our demonstration today was most interesting. It was to some extent a competition between good machinery as represented by the Earthworm tractor and harvester, and poor machinery as represented by the idiotic French contraption which I described in Saturday's report. It was also something of a combination rodeo, Wild West show and Ben Hur chariot race. But the outstanding feature of the whole affair was the highly dramatic conflict between glorious and resplendent virtue as represented by Gadget and me, and foul and slimy vice as represented by the sinister Monsieur Jean Jacques Leboutellier. I will give you all the details of this exciting day in order that you may shudder, as I shuddered, at the villainous machinations perpetrated by the forces of darkness, and then thrill, as I thrilled, at the triumph of honor and decency.

Gadget and I had an early breakfast and arrived at the Château de Moeqethon about seven o'clock. Fifteen minutes later we were out at the field ready to start. About this time the proprietor, Monsieur Cru, appeared and asked us to wait until Monsieur Leboutellier was ready. Monsieur Cru said he wished us to start at the same time, so that he could more easily compare our work. He further explained that the various prominent farmers and other guests he had invited for the occasion would not arrive before half-past eight or nine, and he would prefer to delay the start of the demonstration until they were present. Gadget said that this plan would suit us perfectly, so we left our tractor and harvester standing at the edge of the field and strolled back to the barns to see what was detaining the French harvester man. It was most fortunate that we did this, instead of remaining out in the field, because when we reached the barns we were treated to as interesting and amusing an animal show as anything I have ever paid good money to see in a circus. And this show was absolutely free.

Monsieur Leboutellier had gathered together—partly from Monsieur Cru's farm and partly from other farms in the neighborhood—no less than thirty horses, which he was attempting to hitch onto the front of his huge and clumsy harvester. When I say there were thirty horses, that is exactly what I mean. Gadget and I counted them. They were all splendid, buxom creatures of the Percheron type—large and heavily built, but extremely lively and full of spirit. To control this mass of horseflesh there were, in addition to Monsieur Leboutellier and young André, fifteen drivers—making seventeen men in all. But seventeen were not enough.

The basic trouble seemed to be that all the horses were individualists. They had been accustomed to working, in the time-honored French fashion, on one-horse plows and one-horse carts. Most of them, of course, had had some experience working in pairs, and it is even remotely possible that a few may have been hitched up in a three- or even four-horse team. But any such vast assemblage of draft horses as this was entirely outside their experience. They didn't like it, and they made no attempt to conceal their feelings.

The first part of the hitching up, however, was not so bad. Under Monsieur Leboutellier's direction one pair was hooked to the tongue of the harvester, with another pair beside them. This made four wheel horses, standing side by side. It was evidently Monsieur Leboutellier's plan to have seven ranks of four horses each, with two in the front as leaders. The second group of four horses was harder to get lined up than the first four, and the third and fourth groups brought constantly increasing difficulties.

Monsieur Leboutellier apparently knew nothing of the continuous trace system as used in the artillery for hitching horses in tandem formation. He had worked out, in his ignorance and stupidity, a complicated arrangement of chains, two-horse eveners, and whiffletrees, which added considerably to his troubles. The horses kept backing over these eveners and whiffletrees, and getting their feet all wound up in the traces and chains. The only thing that kept them from getting into a completely hopeless tangle was the fact that Monsieur Leboutellier had provided a driver to ride on the back of the high horse of each pair. These drivers were good French peasants. And although they were apparently as mystified as the horses by this whole weird proceeding, they understood their animals and were able to hold them somewhere near in place. But they could not keep the horses in the rear from playfully nipping at the rumps of the horses in front, and they could not prevent the horses in front from defending themselves with a few swift and well-placed kicks.

There was one horse in particular that made more trouble than any of the rest. He was a magnificent specimen, almost seventeen hands high, pretty near a ton in weight, and glossy black in color. His name was Jacques Johnson, and he seemed to have all the vigor and fighting ability of his distinguished namesake. At first they hitched him up beside the tongue, where he pranced about harmlessly enough for a number of minutes. But by the time they had six teams of four horses each lined up and more or less securely anchored to the various whiffletrees he decided that he needed a little more excitement. He made a quick lunge, and with a vicious snap of his powerful teeth he bit a large wad of hair and a certain amount of hide out of the off hind leg of the animal in front of him. The poor victim let out a startled squeal and launched a terrific kick in the direction from which the attack had come, but good old Jacques Johnson had already dodged back out of harm's way. This little performance of course excited all the other horses; they began prancing, jumping, snorting and wheeling around to such an extent that it took at least ten minutes to quiet them.

It was then decided that Jacques Johnson was too free with his teeth to be trusted behind any of the other horses. Accordingly he and his teammate—a somewhat smaller black horse by the name of Siki— were taken off the tongue and stood at one side to be used as the lead team.

The work of hitching up the rest of the horses proceeded slowly and painfully with much shouting of directions by Monsieur Leboutellier, a great outpouring of quaint French phrases by the drivers and a lot of joyful frisking and leaping about by the horses.

All this time the spectators whom Monsieur Cru had invited were arriving, and they enjoyed the little horse show every bit as much as Gadget and I. Finally, about ten o'clock, the last reluctant animal was jerked into place, and the great French harvester, pulled by the thirty beautiful steeds, with Siki and Jacques Johnson in the lead, moved majestically out to the edge of the wheat field. Here a halt was made while Monsieur Cru explained that he wanted me to harvest the south side of the field, while the French machine was to take the north side. Monsieur Leboutellier walked up to his apparatus, and by pulling a large lever threw in a clumsy-looking jaw clutch which connected the threshing machinery to the large rear wheels. He then announced that he was ready to go.

Gadget and I walked over to our own outfit. I cranked up the tractor and started the motor on the harvester. Then Gadget climbed into the tractor, I took my place in the operator's seat of the harvester, and we started off. The wheat was wonderful—better than anything I have ever seen in America. It was very tall, very thick and loaded with grain. But the machine handled it beautifully, and we proceeded smoothly up to the far end of the field, and then returned. At the end of this first round we noticed that Monsieur Leboutellier's machine had not yet started; so we stopped and walked over to see what was the trouble. Monsieur Cru was kind enough to explain matters to Gadget.

"That large black horse," he said, "has been making trouble again."

"Jacques Johnson?" asked Gadget.

"Yes. That is his name. Just as they were ready to start he kicked the horse behind him with great violence. Apparently he did no serious damage, but as long as he is acting in this way it seems unsafe to leave him in his present position."

"So what are they doing about it?"

"They have sent to the barn for an extra piece of chain. When it comes they are going to hitch Jacques Johnson and Siki two or three meters farther forward. They will then be so far out ahead of the other horses that their kicking can do no damage.

"Well," said Gadget, "I wish them luck. And I suppose we might as well go back to our work. If you want to see some real harvesting you had better come over and watch us."

Note: The above conversation was, of course, all in French. I have given it to you as Gadget translated it to me.

Gadget and I walked back and started another round, and everything went very smoothly. The only trouble was that we didn't seem to be making much impression on our audience. The prominent farmers—to the number of a couple of hundred—were right there where they could have looked at us, but they were so fascinated by the lively animal act which Monsieur Leboutellier was putting on that they seemed to be paying very little attention to us.

We kept on, however, making round after round with relentless regularity. I was using the grain-sacking attachment, and at the end of each round I always had a number of sacks full of grain, which I dumped at the edge of the field.

Toward the end of the morning Monsieur Cru and a few of the farmers came over and watched our work, and at noon Monsieur Cru very kindly invited Gadget and me to have luncheon with him. We accepted the invitation, and after parking the tractor and harvester beside Monsieur Leboutellier's machine, we strolled back to the Château with Monsieur Cru.

"I had hoped," he said, "that Monsieur Leboutellier could join us at luncheon, but he says he will be busy with his horses all through the noon hour."

"I notice that he hasn't got started yet," said Gadget. "What seems to be the trouble now?"

"During the last hour he has had to replace two or three broken whiffletrees, and he has also spent a good deal of time rearranging the teams."

"What was the idea of that?"

"He has been trying to fix it so that all the kicking horses would be at the rear, and all the horses that like to bite would be at the front."

"That sounds like a good scheme," said Gadget. "How does it work?"

"Fairly well," said Monsieur Cru, "except that all of the horses bite to some extent, and all of them kick. I fear that Monsieur Leboutellier has almost more horses than he can manage effectively. But we must give him time to get this thing worked out, and this afternoon I trust we shall see his machine in action."

"I hope so too," said Gadget. "I would hate to see the demonstration too one-sided."

We continued our walk to the Château de Moequethon. Monsieur Cru provided a most excellent luncheon, which lasted until about two o'clock, at which time we drove back to the field in my little rented French car. By this time Monsieur Leboutellier had managed to get all the broken whiffletrees replaced and had succeeded in getting the horses fed, watered and hitched up again. He had found a number of muzzles and had put them on the horses which were the most active with their teeth. He announced that he was ready, so we all stood around expectantly and waited to see him start his monster machine.

With great dignity he climbed up a short ladder and took his place in the midst of a lot of intricate levers and control wheels on the forward deck of his craft. Little André mounted to the poop deck and stood at attention beside another group of levers and wheels. It was very impressive. That French harvester seemed to have more wheels, handles, levers, cranks, adjusting screws and other gimracks than a battleship. Out in front pranced the thirty horses, while the fifteen drivers sat their saddles with grim determination. Monsieur Leboutellier raised his hand. There was a moment of awe-inspiring silence. Then Monsieur Leboutellier's voice rang out loud and clear.

"*Allez!*" he yelled. "*Allez!*"

The drivers took up the cry. "*Allez! Allez!*" they repeated. They leaned forward. They dug their heels into the horses' flanks. But nothing in particular happened. The horses jumped around perhaps a trifle more vigorously than before, but they did not move forward.

Monsieur Cru, who stood near us, diagnosed the trouble at once. "Those horses," he said, "are accustomed to working alone or in pairs. Not one of them will start to pull unless he sees the field clear ahead of him."

"They dragged the thing out here from the barn," said Gadget.

"That was easy. The machine was out of gear. But now that it is in gear the horses will have to pull with all their strength to get it started. And each horse is waiting until the animal in front gets out of his way. None of them, except the leaders, have the slightest intention of moving off."

"As far as that goes," said Gadget, "it doesn't look to me as if the leaders were doing very much either."

We looked at Jacques Johnson and Siki. They stood out in front with their ears back, their noses curled up in a very ugly way, and their tails switching viciously. Their feet were planted solidly and stubbornly on the ground. In other words, they were feeling slightly ugly and distinctly balky. They had made up their minds they were not going to start, and they glared around as if they were daring anybody to make them. It was obvious to everybody that the other twenty-eight horses would never get under way as long as those two balky brutes were out in front of them.

Monsieur Leboutellier climbed down from his lofty perch. He ordered Jacques Johnson and Siki to be unhitched and equipped with muzzles taken from a couple of the other animals. Two of the wheel horses were then unhooked and placed out in front, and Jacques Johnson and Siki were put in their old places beside the tongue.

At this point a most unfortunate accident occurred. Little André—who is a very nice kid and always wants to be helpful—had climbed down from the harvester and had come forward to assist in changing the horses. While he was fastening up one of the traces Jacques Johnson suddenly kicked out behind, and one of his iron-shod hoofs hit poor André in the head. It was not a particularly hard blow, but it knocked him out cold, and he fell down right behind Jacques Johnson, where he was in great danger of being trampled upon or kicked again.

Gadget and I, who were standing nearer than any of the other spectators, immediately ran forward, grabbed André by the feet and dragged him out of harm's way.

"Take him back to the Château," said Monsieur Cru.

I lifted André into the rear seat of my little car. One of the spectators stepped up and said that he was a doctor. We had him get in beside André. Monsieur Cru and I climbed into the front seat.

"Gadget," I said, "you had better stay here with the machine. If you want to cut any more grain before I get back, perhaps you could get one of these French gentlemen to ride on the harvester."

I then drove rapidly to the Château and we took André upstairs and put him to bed.

While the doctor was examining his head he came to, looked around and smiled somewhat feebly. The doctor, after he had finished his examination, spoke briefly to Monsieur Cru, and then repeated his remarks to me in English. Most of these French professional men seem to be pretty good linguists.

"Apparently it is not serious," he said, "but it will be wise to keep him quiet for the rest of the afternoon."

André, however, did not seem to want to stay quiet. He asked the doctor various questions. The doctor replied. Then André looked at me and started a long harangue, which—as it was in the French language—I did not understand. The doctor tried to quiet him, but he kept right on. Finally the doctor turned to me.

"This boy," he said, "has just related a very strange tale, which he wished me to pass on to you."

"Very good," I said. "I am listening."

"First of all," said the doctor, "he wishes me to thank you for saving his life."

"And what gave him the idea," I asked, "that I did any such thing?"

"He asked me exactly what happened, so I told him that after he was kicked you and your wife, at great risk to yourselves, pulled him out from beneath the hoofs of the horses."

"The risk was much less than you think," I said. "I have great respect for the hoofs of that old Jacques Johnson, and I can assure you I kept myself well out of the danger zone. All I had to do was reach in one hand for the fraction of a second and give a strong pull on this young gentleman's left leg."

"Well, be that as it may," said the doctor, "the young man is grateful, and he has told me something which was already on his conscience, and which he would have told before, except that he was afraid of his master, Monsieur Leboutellier."

"Well, well," I said, "this all sounds most amusing. Exactly what is this strange tale?"

"He says that this Leboutellier—seeing that he couldn't compete with your excellent machine by fair means—has determined to get the better of you by treachery."

"The dirty bum!" I said.

"And this noon, while you were absent at luncheon, he tampered with your machine."

"What is that you say? What did he do?"

"The young man's explanation is rather complicated. It appears that Monsieur Leboutellier waited until the drivers had left to take their horses to water. Then he forced our young friend here to help him drain all of the essence from the tractor tank into a number of pails. Or at least that is the way I understood it. That would be perfectly possible, would it not?"

"Yes," I said. "The gasoline, or essence, as you call it, is carried in the large tank which you may have noticed just under the rear end of the hood which covers the motor. All he would have to do is open the side of the hood, place his pail under the drain cock in the bottom of the tank and turn it on. But I don't see how any such proceeding would do him any good. We would discover at once that we were out of fuel and send for some more."

"But wait," said the doctor. "Monsieur Leboutellier did more than that."

"Go on," I said.

"He removed the drain cock and also a bushing which was around it. The young man says that this left a hole more than a centimeter in diameter in the bottom of the tank."

"And what did he do next?"

"He plugged up the hole with a piece of wax, and then they poured all the essence back into the tank."

"This is all beyond me," I said. "What did he think he was doing?"

"After this," continued the doctor, "he took the spark plug out of the rear cylinder, and laid it—with the wire still attached—on a cross member of the tractor frame directly under the wax-plugged hole. Is that clear?"

"It is. Go on."

"He then cranked the motor and had this young man watch the spark plug to see if it gave forth a spark in its new location. Apparently it did. Monsieur Leboutellier then soaked a large quantity of cotton waste in essence and left it on top of the spark plug. After this he closed down the hood and went back to his own machine."

"Holy Moses," I said. "If anybody cranks that tractor it will set all that waste on fire."

"Exactly so," agreed the doctor. "And that is not all. Our young patient says it was Monsieur Leboutellier's intention that the heat from the burning waste should melt the wax so that the entire tankful of essence would run out, catch fire and cause such a conflagration that your entire machine would be destroyed."

"Excuse me," I said as I grabbed my hat, "but I must be on my way."

"Just a minute," said the doctor. "Our friend wants to tell you one more thing. He says that these plans have already partially gone wrong."

"How do you mean?" I asked.

"It seems that just before we returned to the field they heard a gentle, splashing noise from the direction of your tractor. They investigated and found that the piece of wax had dropped out of the hole in the tank and all the essence had run out and soaked into the ground. Monsieur Leboutellier was going to plug the hole again and try to get more essence to refill the tank. But we returned before he could do anything, so they had to leave matters as they were."

"Let me see," I said. "We could still have a small fire in the cotton waste. And we could have something else. Good Lord, yes! That tank is probably full of an explosive mixture of gasoline vapor and air. If that waste catches on fire the flames would go right up through the hole in the bottom of the tank and—Pardon me, but I really must be going."

I went out of that old Château like the well-known bat out of hell. Monsieur Cru followed. We leaped into my little car and headed for the wheat field.

As I sped along I could see ahead of me the vast, looming bulk of the French machine, with the mob of horses in front of it. Beside it stood our neat little harvester. In front of the harvester stood the tractor. And in front of the tractor stood Gadget. Apparently she was just getting ready to spin the crank.

I pushed the accelerator right down to the floor, but I couldn't get any more speed out of the little car. I sounded the horn violently, but it was one of those feeble little French affairs operated by a rubber bulb, and it didn't make enough noise to matter.

Far ahead of me I could see Gadget stooping ever. She was evidently taking hold of the crank. On the seat of our harvester sat a Frenchman with black whiskers—evidently one of the spectators who had volunteered his services. Over by the French machine I could see Monsieur Leboutellier walking nervously back and forth. The thirty horses and fifteen drivers were strung out in front. I recognized Jacques Johnson and Siki on either side of the tongue. Apparently the French outfit had been held up by more trouble of some sort, they had not budged from the place where they had been when I had left.

Gradually, but with maddening slowness, I was drawing closer. I began to hope that I might arrive in time. But I did not. When I was still several hundred meters distant, Gadget gave the crank of the tractor a brisk flip. There was a sudden flash from the center of the tractor. There was a loud report, like about half a dozen sticks of dynamite going off. An irregular-shaped object—which I found out later was the entire end of the tank—went sailing through the air and landed with a resounding whack directly on Jacques Johnson's muscular rear end. This spirited animal—who had so recently balked and refused to move an inch—came to life in a most surprising manner. He leaped forward. He dragged the whole tremendous harvesting machine about six feet. It bumped into the other three wheel horses. These brutes—already half scared to death by the explosion—also plunged madly forward. They butted into the horses ahead. These horses started up. The panic spread. And before the unfortunate drivers knew what was going on, the entire thirty animals were galloping madly across the field, dragging the great harvester after them.

I stopped my little car beside Gadget. Monsieur Cru and I leaped out. I was much relieved to find that she was unhurt. The spectators also had escaped. The explosion had been violent, but there had been no flying fragments except the end piece of the tank which had given such a lusty but harmless blow to Jacques Johnson. I gave the tractor a hasty inspection. The small fire in the cotton waste was already out and the only damage was the destruction of the tank.

By this time the French harvester was nearing the far end of the field. The drivers, although they could not stop the horses, or even slow them down, were able to guide them slightly to the left, so that the huge harvester, in its mad career, swung slowly around in a magnificent circle and finally headed straight back toward where we were standing. The visiting farmers were all in a state of the wildest excitement, rushing hither and thither and shouting and waving their arms around as only Frenchmen can. Monsieur Leboutellier was weeping copiously and rubbing his hands together as if he were engaged in washing them. I think this quaint gesture is commonly described as wringing the hands.

As the harvester, in its circular course, drew nearer to us we were all startled to observe that a very curious phenomenon was taking place. From time to time the huge machine seemed to be throwing out fragments of various kinds, much as Vesuvius in eruption belches forth stones and ashes.

"What on earth is going on?" I said.

"I bet I know," said Gadget. "That machine is in gear. Those horses are pulling it at least five times as fast as it is supposed to go. The threshing machinery inside is being driven at least five times too fast. And it can't stand it."

Gadget was right. The unfortunate insides of the machine had been forced into a state of frenzied activity such as the designers had never dreamed of. The cylinder was spinning like a cream separator, the grain augers were whirling like mad, the fans were going like airplane propellers, and the whole mechanism, as it drew nearer, seemed to be making a noise very similar to the big rattler in the foundry at Earthworm City. And all this accelerated motion was, as Gadget had said, more than the machine could stand.

The entire header, with the cutter bar and reel and everything, had been wrenched off at the far end of the field. And now, as the tremendous apparatus with its plunging horses bore down upon us, we observed a lot of screens, bits of torn canvas and other junk being projected through the side. A moment later we were all running madly for the fence to avoid being hit. But the drivers managed to pull the horses sidewise, and instead of running over us, they went charging by and started another lap of their wild race. As they passed, there was a rending crash and a large assortment of cylinder teeth and about half of the concave shot out through the upper deck. And a moment later a mass of kindling wood that had once been the Jackson feeder was spewed out of the mouth of the dog house.

Once more the horses and their clumsy vehicle receded into the distance. It looked as if the poor animals were getting tired. As they swung around the circle for the second time they slowed down more and more.

And when they got back again to where we were standing, the drivers, with much jerking and sawing of the reins, finally managed to bring them to a stop.

The stupendous French harvester was a complete wreck. The remains of the chassis were towed back to the barns. The horses were unhitched and taken away. And a couple of peasants with a wagon were sent out to the field to gather up as many of the fragments as they could find.

I really began to feel rather sorry for Monsieur Leboutellier. His demonstration had been a complete fizzle, his machine was a total wreck and Monsieur Cru—before I had a chance to do anything myself—confronted him with the proof of his criminal action in attempting to destroy our tractor. Poor Leboutellier saw that denial would be useless. He begged for mercy in a most piteous manner, and I finally told him that if he would pay for the damage I would take no action against him. I named a reasonably high figure, and he paid it at once.

I then asked Monsieur Cru if we had done enough demonstrating to suit him, and he said that we had. However, he did not wish to buy anything until he had had an opportunity to think things over, and he said that he would be glad to talk to us tomorrow morning. Gadget and I, therefore, spent an hour or so interviewing the various spectators who had come to see the demonstration. They spoke highly of our machine, but they all said that their farms were too small to justify the purchase of such a large apparatus. Probably they are right. Apparently Gadget and I will have to be satisfied with the sale of this one demonstration machine to Monsieur Cru, final arrangements for which we hope and expect to make tomorrow morning.

Before we left the Château de Moequethon we learned that young André had so far recovered that he had resigned from Monsieur Leboutellier's employment and taken a job with Monsieur Cru.

Late in the afternoon, when we had returned to Château-Thierry, we were fortunate enough to find at a hardware store a tank of just the right size for our tractor. We will take it out tomorrow and install it. And then, as soon as we have sold the tractor and the harvester to Monsieur Cru. I will write you so that you may congratulate us on our success.

Very truly,
ALEXANDER BOTTS.

ALEXANDER BOTTS
EUROPEAN REPRESENTATIVE
FOR THE
EARTHWORM TRACTOR

HÔTEL JEAN-DE-LA-FONTAINE,
CHÂTEAU-THIERRY, FRANCE.
TUESDAY EVENING, JULY 31, 1928.

MR. GILBERT HENDERSON,
EARTHWORM TRACTOR COMPANY,
EARTHWORM CITY, ILLINOIS.

DEAR HENDERSON: Mere words fail utterly to describe the present mental state of Gadget and myself. But perhaps if I tell you what happened this morning, you may be able to form at least some faint conception of the way we feel this evening.

At about eleven a.m., after we had installed the new tank on the tractor, we saw Monsieur Cru.

"I have decided to buy the French harvester," he said; "not the one which was wrecked, but another one just like it which they are shipping up from the factory."

"I don't understand," said Gadget. "If you want a machine that will harvest your grain and save labor—"

"But I don't," said Monsieur Cru. "What do I want with your labor-saving machinery? Labor is so cheap here in France that there is no point in trying to save it."

"But you told us that you would be interested in seeing us demonstrate."

"Perhaps I should have explained more fully in the beginning. You see, I am a motion-picture producer."

"Yes," said Gadget, "we had heard that."

"I am making a picture called *Les Glaneuses—The Female Gleaners*. The scene is laid in the wheat fields of France. It is inspired by Millet's painting, around which we have woven a story of tender sentiment and burning passion. We wish to give the picture an up-to-date flavor by showing something new and spectacular in the way of harvesting methods. For this purpose your machine is almost worthless. All it does is harvest the wheat. But you saw what Monsieur Leboutellier's machine did."

"Yes," said Gadget. "We certainly did."

"My only regret," he said, "is that we did not have the cameras here yesterday. What a spectacle it was! Ah, those magnificent, charging horses! Ah, the awe-inspiring intricacy of that tremendous machine of Monsieur Leboutellier! It was majestic! It was incredible! It was as a gift from heaven to the motion-picture industry of France. In short, I am sorry to disappoint you, but I am not buying your harvester. I am buying the other one. It is all decided."

And apparently it was. Gadget protested and argued for a long time, but it was no use, so at last we brought the tractor and the harvester back here to Château-Thierry, thus bringing to an absurd and idiotic conclusion what had promised to be one of the most successful selling campaigns of your usually successful salesman.

ALEXANDER BOTTS.

THE MODEL HOUSE

ILLUSTRATED BY TONY SARG

EARTHWORM TRACTOR COMPANY
EARTHWORM CITY, ILLINOIS
OFFICE OF THE SALES MANAGER

SEPTEMBER 10, 1928.

MR. ALEXANDER BOTTS,
MARSEILLES, FRANCE.

DEAR BOTTS: Something over a month ago we received your letter reporting the complete failure of your efforts to sell tractors and harvesters to the wheat growers along the Marne in France. Since that time we have heard absolutely nothing from you. Kindly report at once what you are doing.

Last June we wrote you that we did not consider your success in Europe sufficient to justify us in keeping you over there any longer. At that time we told you to sell the tractors you had on hand as rapidly as possible, after which you and Mrs. Botts were to return to America.

Since then you have accomplished nothing of importance, as far as we know, and we want you, therefore, to make arrangements to come home at once. If you have been unable to sell the tractors that you have with you, you will have them crated up and shipped back to this country.

Very sincerely,
GILBERT HENDERSON,
Sales Manager.

ALEXANDER BOTTS
EUROPEAN REPRESENTATIVE
FOR THE
EARTHWORM TRACTOR

GRAND HOTEL MIRAMARE & DE LA VILLE,
GENOA, ITALY.
MONDAY, SEPTEMBER 24, 1928.

MR. GILBERT HENDERSON,
EARTHWORM TRACTOR COMPANY,
EARTHWORM CITY, ILLINOIS.

DEAR HENDERSON: Your letter has been forwarded from Marseilles. I have not written you because I have been too busy. Besides, I haven't been selling anything. And as I was saying to my wife only last week, "Gadget," I said, "as long as we haven't sold anything, we have nothing to tell, and if we have nothing to tell, what sense is there to writing letters?"

However, as you seem to want to know what we have been doing, I may say that Gadget and I have been making a grand tour through the eastern part of France. After our failure at Château-Thierry, we put the harvester in storage, and, starting on September first, drove the tractor down through Châlons to Chaumont and Dijon, giving demonstrations all along the way in the hope that we might be able to make a sale. From Dijon we continued to Lyons and Grenoble. And all this time we failed to find a prospect that was even warm.

The basic trouble seemed to be the same that we had run into when we first arrived in Europe last spring. With our tractor so expensive, on account of the duty and the freight from America, and with labor so cheap on account of the low wages which prevail over here, they can hire their work done by hand a whole lot cheaper than they can do it with tractors.

At Grenoble we got so disgusted at our failure to make any impression on the French that we shipped the tractor over the Alps by way of the familiar Mont Cenis tunnel, and brought it down here to Genoa in hopes that we might have better luck in Italy.

We arrived yesterday, and at once looked up Marco Manzione, the young Italian whom we had hired on our visit to Italy last spring, and whom we had left here to stir up as much business as possible for us.

Marco had been working hard, but, like ourselves, he was much discouraged. We had left three tractors with him in Genoa, of which he had sold none at all. It seemed to be the same story. The price we had to charge was too much to compete with the low cost of human labor.

Marco, however, had just discovered what seems like a swell prospect for a small machine down in Florence, so he and Gadget and I are taking a ten-horsepower tractor and a plow, and going down there tomorrow.

I note from your letter that you want me to come home at once, bringing with me all my unsold tractors, which number four, all of them here in Italy. It is my intention to follow out your orders exactly. However, Marco states that this prospect in Florence is so good that the sale may be considered as practically made already, so we will stay just long enough to complete the arrangements and will then return home with the three machines.

Most sincerely,
ALEXANDER BOTTS.

ALEXANDER BOTTS
EUROPEAN REPRESENTATIVE
FOR THE
EARTHWORM TRACTOR

HOTEL MINERVA, FLORENCE.
SEPTEMBER 25, 1928.

MR. GILBERT HENDERSON,
EARTHWORM TRACTOR COMPANY,
EARTHWORM CITY, ILLINOIS.

DEAR HENDERSON: Gadget and Marco and I arrived in Florence early this afternoon. Marco had sent word ahead, and our prospect was here at the hotel when we arrived. It is most fortunate that we came, as I am practically sure that this bozo is going to buy a tractor.

His name is Signor Taddeo Ghini. And in spite of the fact that he is an Italian—he was born at a little town called Sanzo, about ten kilometers from here—he is a tall blond with light hair and blue eyes. He seems to have plenty of brains and lots of money. He speaks English very well, as he went to America twenty years ago at the age of fifteen, and has lived there ever since, gradually building up a very prosperous business as a building contractor in a town in New Jersey. And he seems very much interested in the Earthworm tractor. Take it all around, he is obviously a fine fellow, and one whose acquaintance I have a feeling may well be cultivated.

"I am in something of a hurry," he said, when we had introduced ourselves. "I am driving out to look over a farm I have recently purchased near Fiesole. Could you people come along? If I buy a tractor, it would be used on this farm, so it wouldn't hurt you to look it over."

"We would be delighted to come," I said.

Signor Ghini led us out into the Piazza Santa Maria Novella and we all got into his automobile—a very expensive-looking, American, sport touring car. The chauffeur started the motor, and as we drove along Signor Ghini explained to Gadget and to me why he was interested in a tractor.

"I am thinking of buying one of your machines," he said, "as a gift for my father and mother."

"You are very generous," said Gadget.

"I'm afraid I'm not," he said. "Most of my life I have treated my parents pretty badly. I went to America when I was a boy, and stayed over there for twenty years, working hard and making money for myself, and forgetting all about the old folks back here. You know how it is. When you are away from people you forget about them. If you don't watch yourself you lose touch with them completely."

"Yes," I said, "I suppose that is natural."

"It is natural," he said, "but it's not right. Well, one evening last spring I was reading a book."

"You were reading a book?" said Gadget.

Signor Ghini smiled. He had a very pleasant smile. "Yes," he said. "Of course I am nothing but a building contractor, but in spite of that I do read a book once in a while."

"Nothing to be ashamed of," I said. "Even a tractor salesman has been known at rare intervals to do the same."

"This book which I was reading," he continued, "was the *Autobiography of Benvenuto Cellini.*"

"I have read it," said Gadget, who is always right there on this highbrow stuff.

"Maybe you remember, then, madam," continued Signor Ghini, "that this Benvenuto was a pretty wild proposition."

"Yes," said Gadget.

"He was a good artist, and handy with tools, but he was crazy as a bedbug. Anybody he didn't like, he would stick a knife in them, or crack them over the head as quick as you or I would swat a fly. Half his life he spent fighting, for no reason at all except that he had such a vile temper. And yet this man—who had practically no morals at all—was constantly sending money to his old father here in Florence. As I read that book I became more and more ashamed of myself. Here was this wicked Benvenuto, a self-confessed murderer, eking out a difficult and uncertain living at his goldsmithing, and yet sending generous amounts to his father. And there was I, smug, self-satisfied, a supposedly respectable citizen making more money in the house-building business than I could possibly use, and yet keeping it all for myself. You wouldn't believe it, but I had never sent my parents a single cent. And for several years I hadn't even written them. It started me thinking. And when I start thinking, I usually do something about it."

"And what did you do?" I asked.

"The first thing I did, I arranged my business so I could take a long vacation. Then, last June, I came over here and visited my parents. They are very fine people, Mr. Botts."

"I have no doubt of it," I said.

"I had come," continued Signor Ghini, "with the idea of doing something rather substantial for my parents. I am not exactly a multimillionaire, but I have been successful. And as I am not married, I have more money than I need for myself. I had thought of buying my father and mother a large castle or villa somewhere around here, so that they could spend the rest of their days in luxury and elegance. But as soon as I had spent one afternoon with them I saw that this would not do at all."

"Why not?" asked Gadget.

"Because they are plain country people—peasants, I suppose you would call them. They have always lived very simply, and they are too old to change."

"Are you sure about that?" I asked. "I doubt if anybody ever gets too old to appreciate a little extra luxury."

"You are right," said Signor Ghini. "A little extra luxury is all right, but not too much. I asked my parents one day, in an offhand manner, how they would like to live in a big house with lots of servants, and they both laughed derisively at the idea. They have been used to doing their own work, and a lot of servants would only be a nuisance. They absolutely don't want to have anything to do with servants. At the same time, as they grow older, the housework and the chores become more and more of a burden. Right now my younger sister is with them, and she is a great help. But she is going to be married pretty soon, and my parents will be left alone. It will be a little hard for them, but I think I have figured out a scheme that will help them."

"And what is the scheme?" asked Gadget.

Signor Ghini smiled happily. "I think," he said, "I have hit on exactly the thing. About two months ago I bought a small farm with a very simple little house on it, and since then I have worked like mad fixing it up as a model servantless house."

"What a marvelous idea," said Gadget.

"Do you really think so?"

"Of course."

"And do you think my parents will like it?"

"Of course they will."

"I certainly hope so," said Signor Ghini. "I haven't told them anything about it. It is going to be a surprise. I'm going to bring them over tomorrow and present it to them. I am giving the place a last inspection this afternoon. We are almost there now."

All this time the car had been climbing up onto the high land northeast of the city. We passed through a town called Fiesole and finally stopped in front of a small stone house a short distance beyond.

"Come on in," said Signor Ghini.

We entered. We looked around. And for several minutes we were speechless with astonishment and admiration. It was just a little place, but everything in it was absolutely perfect. There was a small living and dining room with stone walls, oak beams overhead, a fireplace, and ancient leaded-glass windows. The furniture was plain and simple. It must have been very old. The wood looked as if it had been waxed and polished for hundreds of years, and it had a rich, warm glow about it that you never find in new furniture.

There were two cute bedrooms, each with a bathroom. At one side of the house was a cool little enclosed garden, and a terrace with a view of the valley of the Arno, the whole city of Florence, and the hills beyond. We looked at this view for at least ten minutes. Then Signor Ghini took us back into the house and showed us the kitchen.

"In furnishing this place," he explained, "I have tried to combine the best of Italy with the best of America. The living room and the bedrooms are in the old Italian-peasant style. The bathrooms and the kitchen are pure American. I am especially proud of this kitchen."

"I should think you might be," said Gadget. "What a kitchen! What a kitchen!"

Signor Ghini was much pleased. "Practically all this stuff came from America," he said. "Look, the electric refrigerator was turned on only this morning, and already we have plenty of ice cubes. This little door is where you throw the garbage. It drops down into the incinerator in the cellar. This other door is the clothes chute."

"And what's this over here?" asked Gadget.

"That's the dishwasher sink. You stack the dishes in this wire basket right over the little paddle wheel. You turn this handle, which lets in hot water. You throw in washing powder. You close the lid. And you start the paddle wheel, which shoots the water all over the dishes. Then you drain out the dirty water, rinse the dishes and the job is done."

"Alexander," said Gadget, "when we get back to California I must have one of these things."

"Over here," said Signor Ghini, "is the electric mixing machine; it mixes dough, grinds meat and coffee, slices vegetables, beats eggs, whips cream, turns the ice-cream freezer and even grinds knives and scissors."

"Swell!" said Gadget. "I must have one of these too."

"The two kitchen cabinets," continued Signor Ghini, "are the best that can be bought."

"And the arrangement is so good," said Gadget. "Everything is placed so you don't have to waste steps at all."

"This thing here," said Signor Ghini, "is the electric wine press."

"Is that American too?" I asked.

"Certainly," he said, "although over there it is sold as a grape-juice extractor. And now we will look at the cellar."

We descended the stairs and inspected the washing machine. It had one compartment for washing, and another where the clothes could be whirled around until dry. Signor Ghini had put in a few old towels for demonstration purposes. He ran some water over them, gave them a spin and then showed us with great pride how the water had been whirled out of them. After this he had us look over the ironing machine, the garbage incinerator, the hot-water heater and the oil-burning, steam-heating plant. Then we went out to the barn.

"My father," explained Signor Ghini, "would never be happy without a little farm work. There are about ten hectares of ground here, and as he will probably want to work it all, I think he ought to have a tractor."

"I am sure of it," I said.

"I have practically decided to get one," he went on, "but I don't know as much about agricultural machinery as I do about household appliances. I hate to bother you too much, but I was wondering if you could bring your tractor out here tomorrow and demonstrate it. I don't want to buy it until I am sure that it will work out all right on this particular farm."

"You are absolutely right," I said. "We will go back to Florence at once, stop at the freight station and get the machine which we have had shipped down from Genoa. We'll grease it up and get it all ready this afternoon. And tomorrow morning we'll bring it out."

"Fine," said Signor Ghini. "I will drive you back to Florence at once."

As we were leaving, a very pretty Italian girl and a nice-looking young Italian man appeared, and Signor Ghini introduced them as his sister and her boyfriend. After chatting with them a few minutes our host led us back to his car and we started along the road to Florence.

"My sister and her future husband," he remarked, "are very keen about this house. They have encouraged me a lot. Sometimes I get scared for fear my parents won't like it, but they tell me I am crazy and that they are sure to just love it."

"Of course they will," said Gadget. "It is a peach of a little place. And it is a wonderful thing you are doing, fixing it up this way and giving it to them."

"I am glad you think so. A lot of people think I am all wrong."

"Applesauce. Why should they think that?"

"There is a lot of criticism by people in the neighborhood—especially those belonging to what you might call the upper crust of society."

"What do they say?" asked Gadget.

"First of all, they say that in Florence and the vicinity there are a great many of the fine old families that were completely impoverished by the war. These people had been used to all the luxuries, and now a lot of them don't even have the necessities. Some of them have to do their own housework. A few have even had to find jobs to support themselves."

"That's tough luck on them," said Gadget, "but what has all this to do with your giving your parents a house?"

"They say that it's an insult to these people—these real ladies and gentlemen—to flaunt in their faces the spectacle of a couple of uneducated peasants enjoying so many luxuries that the real aristocrats now have to do without."

"What else do they say?"

"Oh, they say a lot more. Besides insulting the upper classes, I am starting in to ruin the lower classes. They say that even the best peasants and servants are completely ruined by too much wealth. They become proud and independent. They forget their proper station in life. They become impudent and refuse to work for their superiors except at outrageously high wages. Besides this, they misuse any luxuries you give them. They spend their money foolishly, and in the end are much worse off than if they had been left in their natural state of healthy simplicity."

"And these people think that by giving your parents this house you are going to spoil them completely—is that it?"

"Apparently it is," said Signor Ghini.

"If I were you," said Gadget, "I would go right ahead and spoil them as much as possible. I think they'll enjoy it."

"Of course they will," I said.

"Well," said Signor Ghini, "I'm going ahead with it anyway, and I hope it turns out all right."

When we reached Florence we had the chauffeur drop Marco, Gadget and me at the freight station. We claimed the tractor and the plow, and got them all ready for the demonstration. Tomorrow morning we will get under way early, and by tomorrow night I hope to write you that we have made a sale and contributed our bit toward the ruination of Signor Ghini's respected old father and mother.

Most sincerely,
ALEXANDER BOTTS.

ALEXANDER BOTTS
EUROPEAN REPRESENTATIVE
FOR THE
EARTHWORM TRACTOR

HOTEL MINERVA, FLORENCE.
WEDNESDAY EVENING,
SEPTEMBER 26, 1928.

MR. GILBERT HENDERSON,
EARTHWORM TRACTOR COMPANY,
EARTHWORM CITY, ILLINOIS.

DEAR HENDERSON: What a day this has been! And this evening we absolutely don't know where we are at. Last night I had rather laughed at Signor Ghini's fears that he might fail to please his parents or that he might in some mysterious way ruin them. But tonight there is nothing to laugh at at all. For it appears that Signor Ghini has not only ruined his parents; he seems to have annihilated them entirely. Furthermore, our chances of selling the tractor seem to be fairly slim—and all through no fault of mine, or Gadget's, or Marco's.

Early this morning I drove the tractor out of the city, up the hill, through Fiesole and on to the little villa. Gadget rode beside me and Marco occupied the seat on the plow. We arrived about nine o'clock, and found that Signor Ghini's sister and the boyfriend were already there. They were both much excited. Signor Ghini had gone in his car to his old home in the village of Sanzo, about ten kilometers distant over the hill. He was going to bring the old father and mother to their beautiful new house. He was apt to arrive at any moment.

We drove the tractor out past the barn to the edge of the field which was to be plowed. Here we left it and returned to the house, arriving just as Signor Ghini and his parents came in.

The father was as fine a looking old bozo as I have ever seen. He was tall and blond like his son, but he was much more interesting-looking. His face had been toasted so many years in the blazing Italian sun that it was brown and tough and wrinkled like a baked apple. He was lean but powerful, and he strode into the house with an air of dignity and grandeur that would be hard to beat. The mother also was big and strong, and her face was almost as tanned and weather-beaten as that of her husband.

Signor Ghini had told his parents to come prepared to spend the night, so they had brought a large, dilapidated valise and a small basket which held the family cat—a splendid coal-black creature by the name of Mefisto.

We all entered the house and stood around while Signor Ghini made a little speech of presentation.

At first the old patriarch and his wife couldn't understand what it was all about, but when they finally realized that this beautiful place was all their own, their faces lighted up with broad smiles of childlike delight. They grabbed their son and hugged him and kissed him with gratitude, and then did the same to their daughter and their prospective son-in-law. Signor Ghini glowed with pleasure and satisfaction, and we all joined in with congratulations.

"I told you they'd like it," said Gadget, "and they do."

"Yes," said Signor Ghini. "Isn't it wonderful?"

Everybody seemed completely happy, and even Mefisto, the cat, who had been let out of his basket, wandered delightedly about the room, rubbing himself against the furniture and regarding everything with a benign eye.

Signor Ghini started to take the new owners on a tour of inspection of their house, and Gadget and Marco and I slipped out to start up the tractor and get the plowing under way. I had Marco drive the machine, and Gadget and I stood beside the field and watched the work. What a beautiful picture it was! Overhead, the brilliant blue Italian sky. In the distance, the valley of the Arno and the noble city of Florence. A little

"They grabbed their son and hugged him and kissed him with gratitude."

nearer, the church tower, the villas, the cypress trees and the gardens of the ancient town of Fiesole. And in the foreground, our ten-horsepower Earthworm tractor plowing sturdily along, the very picture of efficiency and beauty.

"Look," I said proudly, "we have here the same thing that Signor Ghini has in his house—a combination of all that is loveliest and most beautiful in Italy and in America."

"How true that is!" said Gadget.

After watching the plowing for a while, Gadget and I strolled back to the house to invite the other people to come out. We found, however, that they were all helping prepare dinner in the elegant new kitchen. They invited us to stick around. We did so, and about twelve o'clock there was a fine meal on the table. Mefisto had a special plate of delicacies in the cellar.

The two old people had been doing a lot of helpless puttering around, but none of the real work. As I remarked to Gadget, they seemed to be bewildered by all the machinery and appliances. I began to wonder whether they really liked this newfangled house, or whether they were just pretending, so as not to hurt the feelings of their son. All through the meal they talked and laughed pleasantly enough, but it seemed to me they were not entirely at their ease.

After dinner I relieved Marco, so he could have something to eat. And when he had finished, I turned the machine back to him and rejoined the rest of the people at the house. When I arrived I found that Mother Ghini was getting a mild bawling-out for having thrown the garbage down the clothes chute instead of the incinerator.

"You had better all come out and see the tractor," I said.

"Fine," said Signor Ghini. "Let's go."

He followed me out to the field, and Gadget and the sister and the boyfriend also came along. The parents seemed more interested in the house, however, so we left them behind. As we watched the tractor, I delivered one of my snappiest little sales talks, proving conclusively that the Earthworm was the perfect machine for this farm.

Just as I was bringing my remarks to a close, there came a series of shouts and a very strange and very loud rattling noise from the house. We all rushed back. In the garden we found Father and Mother Ghini looking very much startled. From inside the house there came a crashing and clattering the like of which I had never heard. I ran into the kitchen. The unholy racket seemed to be coming from the dishwasher sink. I leaped across the room and turned off the switch. The uproar stopped.

I lifted the lid and peered inside at a mass of broken dishes. The rest of the company came crowding in, and before long we found out what had happened. It was very simple. As soon as father and mother found themselves alone, they decided to do a little experimental dishwashing. They put a lot of dishes into the machine. But they left out the wire basket, so all the handsome crockery was piled directly on top of the paddle wheel at the bottom. And when the paddle wheel started whirling it was just too bad.

Note: If any of you bozos back there in Earthworm City buy one of these dishwashers—which you are very apt to do, as they are wonderful machines—you had better tell your wives to be sure to use the wire basket. Otherwise it is just like throwing the plates into a high-powered electric fan.

After the wreckage had been cleared away, Gadget and I drifted back to see how the tractor was getting along. It was still plowing smoothly and steadily. "I am awfully sorry," said Gadget, "that they had that accident. It is too bad."

"The guy has money," I said; "he can buy more dishes."

"It isn't that," said Gadget. "Father and Mother Ghini aren't any too well sold on this house as it is. And it wouldn't take many accidents like that to discourage them completely. And, besides, it takes everybody's mind off our tractor demonstration. And that is very bad for us."

"Yes," I agreed. "It is."

"After all the excitement," said Gadget, "it is very possible that we won't be able to get any of them to come out and even look at the tractor."

But she was wrong about that. A few moments later, Signor Ghini, with the sister and the boyfriend, joined us. We were pleased to see that they were still interested in our machine.

"I am sorry you did not bring your father and mother out," said Gadget. "I'd like them to see this demonstration."

"They are so much interested in the house," said Signor Ghini, "that they wanted to stay behind."

"You are not afraid to leave them alone with all that unfamiliar machinery?"

"Oh, no," he said. "They'll be all right."

From the direction of the house we heard loud screams and calls for assistance. Mingled with the human cries were a few frantic yowls that could only have come from Mefisto, the cat.

"Good Lord!" said Gadget. "What have they done now?"

We all ran as fast as we could, but by the time we reached the house the calls for help had ceased. All was quiet. Not a person was to be seen. We entered the kitchen. Nobody was there. We went through the rest of the house, looking in closets, under beds—everywhere. No result. Everything was in order; nothing, apparently, had been disturbed; but there was no sign of the Ghini parents. And there was no trace of Mefisto.

"They called for help," said Signor Ghini. "They were in trouble. They needed assistance. Where are they?"

We inspected the cellar, but found nothing.

"We've got to find them!" said Signor Ghini. "And we've got to hurry! This may be serious!"

We went out into the garden. We peered under the rosebushes. We looked over the terrace. No trace of father. No trace of mother. No trace of Mefisto.

By this time the neighbors had begun to appear. They had heard the bloodcurdling cries for help, and had come to see what was the matter. The first arrival was a little fat man from next door. He had big black mustaches. Behind him came two elderly ladies, a small boy and a couple of dogs, but none of them seemed to be able to do anything but ask foolish questions and rush hither and thither about the garden and add to the general excitement.

Signor Ghini became more and more agitated. "We've got to do something," he kept repeating. "We've got to do something."

But none of us could think of anything much to do. At last a couple of *carabinieri* arrived. Under their direction a number of searching parties were organized. The news of the strange disappearance spread, and there were plenty of volunteer helpers. With a force of at least a hundred people we worked hard all afternoon, beating around through the beautiful gardens, vineyards, fields, roads and lanes on all the slopes about Fiesole. But all without any result whatever. The old couple and the black cat had vanished without leaving a trace.

At six o'clock Gadget and Marco and I returned here to the hotel, leaving the search in the hands of the authorities and the numerous volunteers. We had some thought of going back this evening, but decided that they had plenty of help and that we would not be needed.

I have been spending the evening writing this report. Naturally, there has been no chance for us to discuss tractors with Signor Ghini, so I will have to close without letting you know whether or not we have any chance to make a sale. So no more at present from your tired salesman.

Alexander Botts.

ALEXANDER BOTTS
EUROPEAN REPRESENTATIVE
FOR THE
EARTHWORM TRACTOR

HOTEL MINERVA, FLORENCE, ITALY.
THURSDAY, SEPTEMBER 27, 1928.

MR. GILBERT HENDERSON,
EARTHWORM TRACTOR COMPANY,
EARTHWORM CITY, ILLINOIS.

DEAR HENDERSON: While we were eating breakfast this morning, who should walk in on us but Signor Taddeo Ghini himself.

"Well," he said, as he sank wearily into a chair, "we found them at last."

"Are they all right?" asked Gadget.

"They are," he said.

"What happened? Where are they?"

"It is very disappointing," said Signor Ghini.

"But what happened?"

"It is a bit complicated. You remember Mefisto?"

"The big black cat?" Gadget asked him.

"Yes," said Signor Ghini.

"Did you find him too?" asked Gadget.

"We did."

"And he's all right?"

"He's all right now. But yesterday he had a little accident. Probably you remember that we gave him a big meal down the cellar. When he finished he probably felt a bit sleepy. He began looking for a good place to take a nap. And he discovered those towels that I had left in the drying compartment of the washing machine. So he jumped in. And that is where he made a mistake."

"How so?"

"He is a big, clumsy brute," explained Signor Ghini. "When he leaped into the drying compartment he must have bumped into the hinged cover in such a way that he knocked it shut after him. The cover has a spring catch. So there was Mefisto—inside the machine, with no way to get out."

"But if that's all that happened," said Gadget, "I don't see—"

"That isn't all," said Signor Ghini. "When Mefisto woke up from his nap and found he was a prisoner he began mewing gently but pitifully. My father and mother heard him and went down the cellar to see what was the matter. They tried to open the cover, but they are not used to machinery and they made an error."

"What did they do?"

"They monkeyed around, trying to release the catch. And before they realized what they were doing they had turned on the electric switch and started the dryer rotating."

"With the cat in there?"

"Yes," said Signor Ghini. "Unfortunately for him, Mefisto was in the dryer. And it certainly took him for a ride. What a whirl he must have had."

"Did it kill him?" asked Gadget.

"Oh, no. You can't kill a cat so easy as that. But he didn't like it. It was then that he set up the frantic yowling that we heard away out in the field. My parents shouted for help and worked on the machine as fast and as furious as they could. They finally succeeded in getting the current turned off and the lid opened."

"And they took Mefisto out?"

"They didn't have to. He came out himself. And my parents say that never have they seen that animal so active. He bounded out so vigorously that he almost hit the ceiling. And as soon as he landed on the floor he started traveling."

"Where did he go?"

"At first he didn't go anywhere. He had plenty of strength and energy, but apparently he was so dizzy he couldn't coordinate his movements very well. All he could do was circle around and around on the cellar floor at an amazing speed. But finally he shot up the cellar stairs, did a couple of figure eights in the kitchen, skidded out the door, veered and looped through the shrubbery and finally worked his way around the house and out into the road in front."

"And what were your father and mother doing all this time?" asked Gadget.

"They were following, trying to catch him. But he was too quick and his course was too shifty. By the time he reached the road, he could handle himself a little better. He was still doing ellipses and parabolas from time to time, but his general direction took him straight down the road. And this all happened so fast that Mefisto and my father and mother were some distance away by the time we reached the house. That is why we missed them. And as the road is hidden by stone walls, none of the neighbors saw them either."

"Did they finally catch the poor animal?"

"Yes. Before long he began to get tired and they grabbed him. They sat down under a tree and held him in their arms and petted him. And before long he was all right again. But my father and mother were completely discouraged. They decided not to come back, for fear I would try to persuade them to stay in the new place. They took Mefisto and walked all the way back to their old home at Sanzo. I found them there last night, very much refreshed by their little hike, and tickled to death to get home again. And Mefisto was as happy as they were."

"We are awfully glad to hear it," said Gadget. "Everything is all right again."

"I wouldn't say that," said Signor Ghini. "Of course, it's a great relief to find them safe and sound. But I am a very disappointed man, Mrs. Botts. A very disappointed man."

"How so?"

"My parents have decided that they would never be happy in the new house. They say it would be too much trouble for them to try to learn how to run all the machinery. They have always lived in the old home at Sanzo, and they say they are too old to change."

"So it's all off?" asked Gadget.

"It's all off," he answered sadly. "Perhaps these upper-crust people are right, and it is a mistake to give luxuries to what they call the lower classes."

"Boloney!" said Gadget. "The only trouble with your parents is that they are so old they are set in their ways. Their position in society has nothing to do with it. If they belong to the lower classes, so do you, and so do I, and my husband, and your sister, and that young guy she goes around with. But we all appreciate good things. If you don't believe it, just offer that house to your sister, and see what she says."

"As a matter of fact," said Signor Ghini, "I offered it to her this morning, and she took it. She and her husband will move in next month, right after their wedding. They are both so crazy about the place that I decided it would be better to give it to them than to sell it to strangers."

"Good for you!" said Gadget. "The more I see of you, the more I like you."

"As long as I am giving a present," continued Signor Ghini, "I might as well do it right. So I am going to buy your tractor to go along with the place."

"Splendid!" said Gadget. "My regard for you increases all the time. In my opinion, you are a credit both to Italy and to America."

And I may add that this opinion is concurred in by

Yours truly.

WE'RE GOING TO RUIN THE LOWER CLASSES

ILLUSTRATED BY TONY SARG

EARTHWORM TRACTOR COMPANY
EARTHWORM CITY, ILLINOIS
OFFICE OF THE SALES MANAGER

OCTOBER 9, 1928.

MR. ALEXANDER BOTTS,
MARSEILLES, FRANCE.

DEAR BOTTS: This will acknowledge your letter of September twenty-seventh, reporting the sale of one ten-horsepower Earthworm tractor to Signor Taddeo Ghini in Florence, Italy. We are, naturally, pleased that you put this over, but we must remind you that one sale every two or three months—which seems to be the rate you are now going at—is not enough to justify us in keeping you in Europe any longer.

You have been there seven months—long enough for a fair trial. During this time we have been paying your salary and all expenses for yourself and your wife. The results have not been sufficient to justify this expense.

You will doubtless remember that in a letter dated June tenth and in another letter dated September tenth we suggested that you make every effort to dispose of the tractors you had on hand and arrange your plans so that you and Mrs. Botts could come home in the near future. Apparently you have not yet been able to sell the tractors which you have with you, and, as far as we can tell from reading your reports, you have made no definite plans for returning to America.

As you have done nothing toward following out our suggestions, we now order you definitely to start back within two weeks at the latest.

If you cannot sell your tractors within that time—even by reducing prices—you will have to have them shipped back here.

Very truly,
GILBERT HENDERSON,
Sales Manager.

CABLEGRAM
WEEK END LETTER
ROME ITALY OCT 20 1928
HENDERSON EARTHTRACT EARTHWORM CITY ILL

YOUR LETTER FORWARDED FROM MARSEILLES STOP CANNOT COME RIGHT NOW STOP HAVE BIG NEW IDEA

BOTTS

CABLEGRAM
EARTHWORM CITY ILL OCT 22 1928
ALEXANDER BOTTS ROME ITALY

NEVER MIND BIG IDEA STOP COME HOME

HENDERSON

ALEXANDER BOTTS
EUROPEAN REPRESENTATIVE
FOR THE
EARTHWORM TRACTOR

HOTEL EXCELSIOR, ROME, ITALY.
TUESDAY, OCTOBER 23, 1928.

MR. GILBERT HENDERSON,
EARTHWORM TRACTOR COMPANY,
EARTHWORM CITY, ILLINOIS.

DEAR HENDERSON: Your cablegram arrived last night. At first I thought of wiring back to explain my plans, but I finally decided not to. About all I could have told you in a cable—unless I had made it very long and shockingly expensive—would have been the bare fact that I plan to stay over here quite a while longer. I would not have had the chance to give you all my reasons. And this might have worried you or even irritated you. You might have come to the erroneous conclusion that I was disregarding your instructions. Nothing could be further from the truth. I have always held you in the greatest respect, and I wish to assure you that I have given your letters and your cablegrams the most thoughtful consideration. But it has been impossible for me to take your remarks with any great seriousness because you are ignorant of certain vital facts and considerations which change the whole aspect of the situation. A smaller man, in my position, might have spoiled everything by giving you the stupid, literal type of obedience which is in vogue in the Army. But I prefer to follow the higher principle of discipline by which a subordinate obeys not the order which his superior has actually given him but rather the order which that superior would have given if he had known what he was talking about.

I want you to understand that I am not criticizing you in any way. Considering your ignorance, you acted, after all, in what was a very reasonable manner. I can imagine exactly how you thought the matter out. You probably said to yourself, "Here is old Botts claiming that he has a big idea. Well, he's had big ideas before; and some of them, when he came to work them out, didn't amount to bug dust. So we'll just tell him to forget his schemes and come home." And very likely your reasoning was pretty good.

The only trouble was that you didn't know that this time I had a big idea that is not a little big idea but a very big big idea. It is really stupendous, and it will be a red-letter day for the Earthworm Tractor Company when I get it all worked out and put into action. And right at this point, Henderson, old boy, I want to put in a word of cheer for you personally. When I have achieved my big success over here, you may get to thinking about the fact that you ordered me to come home, and that if I had followed your directions the big opportunity would have been lost. This may have a tendency to make you feel pretty cheap, and maybe apologetic. You might even get the idea that I would be sore at you for attempting to block my plans. So I want to say right here and now that there will be no need for you to feel embarrassed or blame yourself at all. Any one of us is liable to make a mistake once in a while. And don't worry about my attitude. I am not a man to harbor a grudge; especially when I know that this blunder of yours was made through ignorance and not through malice.

And now that I have explained my position in the matter of our business relationship, I will tell you all about my plans. It will be absolutely necessary for my wife and me to leave Italy in about two months. The reasons for this I will explain in due course. For the present, on the other hand, it is absolutely essential that we stay over here. The reasons for this I will explain at once. I have—as I have mentioned before—a new and very large idea. It is, in fact, a regular wow of an idea, and is the result of long-continued and highly concentrated thinking. It began to take form as long ago as September thirtieth. My wife and I—just back from Florence, where we had sold the ten-horsepower Earthworm to Signor Taddeo Ghini—were sitting in our room in the Grand Hotel Miramare & de la Ville at Genoa. We were talking about your letters of June tenth and September tenth.

"In some ways," I said, "Henderson is right. We haven't done as well as I had hoped."

"We have made a certain number of sales," said Gadget.

"Yes," I said. "We sold three tractors to the man on the boat, four tractors and a lot of machinery to the American contractor at Tarascon—"

"That makes seven," said Gadget.

"We failed at Merano," I continued, "but we certainly came through in Venice. It was a big day in my life when we actually succeeded in selling a tractor in that waterlogged old town."

"The one in Venice makes eight," said Gadget.

"We sold six to the Italian Railroad Company, and six to those people in Russia."

"That makes twenty," said Gadget. "We won't say anything about the one that ran over the cliff into the Bug River."

"The less said about that," I remarked, "the better. Then we sold four for Mr. McGinnis up in Germany."

"Which makes twenty-four," said Gadget. "Then one to that Frenchman, and one more down at Florence, making a grand total of twenty-six. That's not so bad."

"In many ways it is pretty good," I said. "But Henderson claims most of our sales have been mere freak stunts. He is discouraged because we haven't opened up any big, steady market. He seems to think conditions over here are hopeless."

"Of course," said Gadget, "conditions really are pretty bad. We have to sell our tractors at almost double the American price. How can we expect people to buy them when they can hire their work done by hand at about a third what it would cost at home?"

"If only we could reduce prices!" I said.

"We can't," said Gadget. "The freight and the duty are too high."

"I wonder," I said, "if there is any chance of wages in Italy going up?"

"I doubt it," said Gadget. "For one thing, all the employers and all the people with money think it is morally wrong to give the poorer classes more than they absolutely have to. That's what disgusts me the most. It's a cockeyed theory. What harm could it possibly do to pay higher wages?"

"You know the arguments," I replied. "If you overpay the lower classes, you just naturally ruin them. The money does them no real good; all they do is spend it. Furthermore, they get exaggerated ideas of their own importance. They become lacking in respect for their superiors. They develop an improper taste for the sort of luxuries that belong by right only to the upper classes. And they are completely ungrateful. They never even thank you for the extra pay. But they become insubordinate and refuse to work if it ever becomes necessary to reduce their pay to a reasonable rate of fifty cents a day or whatever it is."

"Shut up," said Gadget. "It is all a lot of tripe and it makes me sick."

"I know," I said, "and it has exactly the same effect on me. But what can we do about it?"

"Nothing," said Gadget. "I suppose we'll just have to make up our minds to be faintly nauseated throughout the rest of our stay."

"All right," I said. "So let it be."

And for a while it was. We put in a couple of hard and discouraging weeks chasing around looking for somebody to buy our tractors, so that we could wind up our business and come home, according to your instructions. We had no luck at all. And Marco Manzione, the young Italian whom we had hired and who was up at Milan, kept sending us reports that he was not accomplishing anything either. But all this time—although I never suspected it—the beginnings of my big idea were slowly germinating in the deep and hidden chasms of my subconscious mind.

And then, all at once, things began to happen. One morning last week Marco called me up by telephone from Milan. This in itself was most sensational. Because when anybody in Italy actually uses the long-distance telephone, it always indicates that something highly unusual is going on. And this case was no exception. Marco was all in a flutter. He talked very fast. And as the connection was rotten and his English is none too good anyway, it took me quite a while to get what he was driving at. Finally I gathered that one of the Earthworm tractors which we had sold last May to the railroad was working on a dirt-moving job at a place where they were realigning the tracks not far from Milan. It had been doing the work so well that quite a number of people had gone out to observe and admire it.

"And this afternoon," said Marco, his voice quivering with joy, "a wonderful thing will occur."

"All right," I said. "What will occur?"

"Our Earthworm tractor—I mean the one we sold to the railroad—is to be visited and inspected by a member of the cabinet—one of the highest officials in Italy."

"Well, well," I said. "Isn't that nice?"

"It is superb," said Marco. "If we can get this man interested in Earthworm tractors, our fortunes are made."

"Possibly so," I said, "but probably not."

"I assure you," said Marco, "it is the chance of a lifetime. I called to see if you could come up at once. This is such an important occasion that I couldn't take the responsibility of handling it all myself."

"How long does it take to get up there?" I asked.

"Three or four hours, I think."

"All right. Gadget and I will come along on the first train we can catch. Thanks for calling up."

I told Gadget the news. We hurried down to the station. And we took a train that got us to Milan early in the afternoon. We found Marco waiting for us at the station.

"You are too late," he said. "The plans were changed. The cabinet member had to leave for Rome at noon, so he inspected the tractor this morning."

"Were you there?" I asked.

"I was."

"And how did everything go?"

"Not so good, Signor Botts—not so good."

"That's too bad. What happened?"

"The man came. He saw the tractor at work. He talked with the operator. Then he talked to me. And I am pleased to say that he saw at once the possibilities behind the machine."

"What did he say?"

"He said that the Earthworm was far superior to any tractor he had ever seen. He said it would be the perfect machine to use in building our proposed system of grand automobile routes, and that it would be invaluable in the army for moving supplies over rough country and pulling artillery."

"A very sensible thought," I said. "Did you tell him we would be glad to sell him as many Earthworms as he wants?"

"I did."

"What did he say?"

"He asked me if it were true that the machines were made only in America. I said that they were, and he then stated that it would be impossible in that case for the Italian Government to buy any of them."

"Why?" I asked.

"Because Italian industries must be encouraged. It is better, he says, to buy a poor tractor made in Italy than a good one from a foreign land. And that is the tragic part of the whole thing. He admires our Earthworm tractors, but he feels it would be wrong for him to buy them."

"Possibly, from his point of view, he is right," I said. "But it certainly is tough luck on us. You say he has gone back to Rome?"

"Yes."

"Then I guess this little episode is about over. But cheer up. If we can't sell him, we may be able to sell somebody else. And as there doesn't seem to be anything further for me and Gadget to do up here, we will return to Genoa on the next train." A half hour later we were rolling southward, and we got back here just in time for supper. During the entire train ride I kept meditating on the various phases of our European selling campaign. I continued this meditation while we were eating our evening meal. And all this time no ideas of any particular value seemed to emerge. But after we had finished supper and returned to our room, a remarkable phenomenon occurred. Suddenly, with no warning, all the chaotic thoughts and ideas which had been moving aimlessly through my mind for weeks clicked into place and formed, in the twinkling of an eye, a single magnificent and logical conception. I had hit upon the perfect plan for solving all our difficulties.

It was much the same sort of sudden inspiration as came to old Archimedes the day he was sitting in his bathtub, the chief differences being that my idea was probably better than his, and I have more sense than to go rushing about the neighborhood with no clothes on, shouting "Eureka."

I did, however, slap Gadget on the back in a way that somewhat startled her, and I began to tell her all about it.

I will now repeat for your benefit the long, and no doubt rather tedious, conversation which I had with Gadget. I do this, not because my remarks were interesting—they were far too deep and intricate for that—but because if you struggle through them you will get at least a faint grasp of the basic facts concerning the mighty project I am about to launch.

"Congratulate me, Gadget," I said. "I have hit upon a big, new idea. We must start an Earthworm tractor factory over here in Italy."

"That doesn't sound so big to me," said Gadget. "And it's not new. Mr. Henderson told us before we came over that if we worked up a large enough market the company might eventually build a European factory."

"I know," I said. "And he had the thing completely backward. My idea is that if we build the factory first, then we are absolutely sure to work up the market."

"I don't see that at all."

"It was your idea, wasn't it, that the two big factors working against us are the high price of the tractors and the low price of labor?"

"Yes."

"All right. We make the tractors in Italy. We save the freight. We save the duty. We can sell the machines just as cheap as we do in America."

"We can't unless we get a big volume of sales, so the factory can be run on a quantity-production basis."

"All right," I said. "We're going to have a big volume of business."

"How?"

"In the first place, we're going to sell a whole lot of machines to the government. Did not that cabinet member say that he needed our machines for his army and his road building? And did he not say that the only thing which kept him from buying them was the fact that they were built outside of Italy?"

"But even if you do get a big government order," said Gadget, "that won't keep the factory running forever. You will still be competing with your machines against cheap manual labor."

"Absolutely not," I said. "It is on this subject of wages that my scheme rises to really sublime heights."

"I don't see how," said Gadget.

"I am going to play a little trick on the entire Italian nation," I said.

"I still don't understand," said Gadget.

"Well," I said, "it's going to be an imitation of a little trick that was once played in America. Probably you are too young to remember what happened on January 5, 1914."

"I was in school at that time," said Gadget, "but I can't say I remember anything that occurred on that particular day."

"Probably you wouldn't," I said. "But it was, nevertheless, a date of national importance—as worthwhile remembering as October 12, 1492, or July 4, 1776."

"All right," said Gadget. "Go on and tell me what happened, if anything."

"On January 5, 1914," I said, "Mr. Henry Ford announced a minimum wage for all his employees of five dollars a day."

"Is that all?" said Gadget.

"That was enough. At that time the average factory worker thought he was lucky if he got two bucks a day. The news was discussed with bated breath in garages, in factories, in country stores, everywhere. There was one remark that I remember was repeated over and over again. 'They say,' some two-dollar-a-day mechanic would whisper, 'that even the guy that sweeps out the shop gets five dollars a day!' Then some one-dollar-a-day ditch digger would repeat in an awed voice, 'Five dollars a day!' And all the fifteen-dollar-a-week clerks and the thirty-dollar-a-month farm hands would sit around with their mouths wide open in astonishment. That guy with the broom became a national celebrity. People could see him in their imagination—the man with the broom—a simple-minded creature, infinitely lower in the scale of life than the man with the hoe. In his dense ignorance he was incapable of any such intricate job as tightening up Nut No. 64. And yet this hulk of flesh, merely because he worked in the magic Ford factory, pulled down his five good bucks each and every day. It all sounded like a wonderful fairy tale."

"How did it sound to other factory owners?"

"Terrible. They were driven half crazy by all the help coming in, asking for more pay, and trotting out the morning paper where Henry Ford was quoted as saying that the way to make big profits was to raise wages and reduce prices."

"And did these other employers raise wages?"

"They had to, or all their men would have gone to Detroit. In order to pay the extra wages, they had to get more efficiency. To get more efficiency, they had to install labor-saving machinery and go into quantity production. To sell the extra stuff they made, they had to lower prices—"

"And they all went busted?"

"No. That's the funny part of it. In places where they held wages down, they had a hard time. Wherever they raised wages they seemed to make more money than ever. And as soon as everybody had more money, everybody bought more stuff, and that made more business, which made more money. And that's why wages in America are higher than anywhere else."

"And you claim this was all caused by Henry Ford?"

"Maybe," I admitted, "he didn't do so much as it seems he did. But he got in at just the right moment with the heavy publicity. And that's just what I'm going to do over here."

"You mean you actually think you're going to force up wages and ruin the lower classes over here the way Henry Ford did in America?"

"Exactly."

"It seems to me you're taking on a pretty big job."

"I know it. But that only makes it more interesting. The first thing I want to do is get the Italian Government lined up. We'll go down and explain to the officials what a grand thing it would be for everybody if we built an Earthworm tractor factory in Italy. It will give the government cheap tractors. And it will give the people high wages and general prosperity. In return, I'm going to ask for some special favors in the way of advance orders, free port privileges and various other things. With the government on our side, we can't help but succeed."

"I am still a little skeptical," said Gadget, "but it's worth trying, anyway."

"Then you're with me?"

"With you absolutely," said Gadget. "Hooray for the ruin of the lower classes!"

The above conversation I have repeated in full, so that—as I told you before—you could get a complete idea of what we are going to do. And now that you know what a splendid plan we have, you will, no doubt, want to begin preparing for action. A word of caution is, therefore, necessary. It will be all right for you to start planning on just where you will get the vast sums of money which will be needed. But don't actually borrow the capital or send any large number of people over here until I give you the word that everything is settled, because there is always a remote possibility that things might fall through.

I will keep you informed as to the exact progress of affairs. Gadget and I have come down here to Rome. With the help of an Italian lawyer we have drawn up an agreement between the Earthworm Tractor Company and the Italian Government. It is very carefully written and covers all points with great clarity and thoroughness. We have requested an interview so we can present this document to some of the big bugs in the government. We'll let you know how we come out.

Most sincerely,
ALEXANDER BOTTS.

CABLEGRAM

EARTHWORM CITY ILL NOV 5 1928
ALEXANDER BOTTS ROME ITALY

YOUR LETTER RECEIVED STOP EUROPEAN FACTORY IMPOSSIBLE AT THIS TIME STOP COME HOME AT ONCE

GILBERT HENDERSON

CABLEGRAM

ROME ITALY NOV 6 1928
GILBERT HENDERSON EARTHTRACT
EARTHWORM CITY ILL

HADNT I BETTER WAIT UNTIL I SEE GOVERNMENT OFFICIALS QUESTION MARK

BOTTS

CABLEGRAM

EARTHWORM CITY ILL NOV 7 1928
ALEXANDER BOTTS ROME ITALY

NEVER MIND OFFICIALS STOP COME HOME AT ONCE

HENDERSON

CABLEGRAM

ROME ITALY NOV 8 1928
GILBERT HENDERSON EARTHTRACT
EARTHWORM CITY ILL

HAVE DECIDED TO STAY A WHILE LONGER

BOTTS

CABLEGRAM

EARTHWORM CITY ILL NOV 9 1928
ALEXANDER BOTTS ROME ITALY

CANNOT BACK YOU ANY LONGER STOP HAVE STOPPED YOUR SALARY AND EXPENSE CHECKS

HENDERSON

CABLEGRAM

ROME ITALY NOV 10 1928
GILBERT HENDERSON EARTHTRACT
EARTHWORM CITY ILL

SAW ONE OF THE BIG BRASS HATS TODAY STOP GAVE HIM TENTATIVE WRITTEN AGREEMENT STOP HE PROMISED TO CONSIDER IT AND LET US KNOW HIS DECISION IN A FEW WEEKS STOP I AM AMAZED AT YOUR CABLEGRAM STOP MY EXPENSES HEAVY HOTEL AND CABLE BILLS VERY HIGH AM ALMOST OUT OF FUNDS NOT ENOUGH LEFT EVEN TO BUY PASSAGE HOME STOP YOU MUST CONTINUE SENDING CHECKS

BOTTS

CABLEGRAM

EARTHWORM CITY ILL NOV 12 1928
ALEXANDER BOTTS ROME ITALY

WE WILL SEND YOU NO MORE MONEY STOP BUT WE DONT WANT YOU MAROONED OVER THERE SO WE ARE MAILING YOU TWO TICKETS HOME TO NEW YORK

HENDERSON

ALEXANDER BOTTS
EUROPEAN REPRESENTATIVE
FOR THE
EARTHWORM TRACTOR

HOTEL EXCELSIOR, ROME, ITALY.
NOVEMBER 14, 1928.

MR. GILBERT HENDERSON,
EARTHWORM TRACTOR COMPANY,
EARTHWORM CITY, ILLINOIS.

DEAR HENDERSON: If I did not have such a sweet disposition I would be sore as a boil. If I were to treat you as you deserve, I would resign and leave you flat. But my loyalty to the company is such that I am going to carry on a little longer.

I have tried to explain to you how important it is that I should remain here to close up this deal with the Italian Government. Apparently you are too dumb to understand.

Very well; I won't argue with you. I will just stay on.

It is true that our funds are shockingly low, but we are far from helpless. I have paid Marco's salary up to the end of the month, so he won't rate anything more until December thirty-first. Gadget and I are moving from this expensive hotel today and taking a cheap light-housekeeping apartment down near the river.

And with my usual enterprise and resourcefulness I have got myself a job. It was yesterday morning that I heard of a vacancy on the staff of one of the local tourist agencies. I spent the whole day, with Gadget's assistance, studying histories and guidebooks of the city of Rome. And late in the afternoon I descended on this agency with such a fluent and amazing line of talk that they at once hired me as a lecturer and guide on one of these rubberneck wagons that takes Americans around to see the sights. I start work tomorrow morning. The pay is small, but it will be enough.

So you can keep your money. I won't demean myself by asking for any more until I have put through this deal. And I don't care whether you mail me those tickets or not. I can assure you that I shall have absolutely no use for them.

ALEXANDER BOTTS.

ALEXANDER BOTTS
EUROPEAN REPRESENTATIVE
FOR THE
EARTHWORM TRACTOR

ROME, ITALY.
NOVEMBER 26, 1928.

MR. GILBERT HENDERSON,
EARTHWORM TRACTOR COMPANY,
EARTHWORM CITY, ILLINOIS.

DEAR HENDERSON: I called at the Excelsior Hotel this morning and was much pleased to find that the two steamship tickets had arrived. It was most kind of you to send these over, and both Gadget and I want to thank you. In my letter to you a couple of weeks ago I believe I remarked in a rather high and mighty manner that I did not care whether you sent these tickets or not, and that I would have absolutely no use for them. Well, it turns out that I was wrong. We shall have to use them after all.

The reason for this change in our plans is that things are not going as well as I had expected. In the first place, the Italian Government is taking longer than I had expected to consider my proposition. It appears that I cannot hope to get their answer for another week or two.

In the second place, my job with the tourist agency was very disappointing. On the first morning, when I reported for work, the manager asked me to accompany a party which was going to visit St. Peter's and the Vatican. This pleased me very much.

With the help of Gadget and the guidebooks I had worked up a very brilliant and impressive discourse about this great basilica and the famous residence of the Popes.

Unfortunately, however, I inadvertently got on the wrong bus, and it was not until after I had talked and rhapsodized for a half hour or more that I discovered I was giving the St. Peter's lecture in a church known as St. John Lateran.

When we got back from the ride the cash customers made a great uproar about my mistake, and the manager was foolish enough to fire me at once. I explained to him that there was nothing wrong with my lecture, except that it was given in the wrong place, and I pointed out that my mistake was very natural, in view of the fact that I had never before visited either of these churches. Unfortunately, however, he was as unreasonable

"When we got back from the ride the cash customers made a great uproar about my mistake, and the manager was foolish enough to fire me at once."

as some of the executives of the Earthworm Tractor Company, and he continued to insist that he wanted nothing further to do with me. In the end I saw that further argument was useless, and gracefully withdrew.

After this I spent day after day tramping the streets looking for work. But it seems almost impossible for an American to pick up a job in Rome. At any rate, I had no luck. And all the time our small supply of cash was steadily wasting away. Last night we spent our last lira, and our situation was becoming desperate.

But this morning, when the tickets arrived, they brought new hope. Gadget and I at once decided to swallow our pride and use them. Accordingly, I took them down to the steamship office and explained to the man in charge that I wanted to turn them in. He was very polite and allowed me the full value minus a very small cancellation fee. This gives us ample funds to live over here for several months longer, if necessary.

We are most grateful to you, Henderson, for getting us such expensive tickets, and for neglecting to have them restricted to our personal use only.

Yours appreciatively,
Alexander Botts.

ALEXANDER BOTTS
EUROPEAN REPRESENTATIVE
FOR THE
EARTHWORM TRACTOR

ROME, ITALY.
DECEMBER 8, 1928.

MR. GILBERT HENDERSON,
EARTHWORM TRACTOR COMPANY,
EARTHWORM CITY, ILLINOIS.

DEAR HENDERSON: We are still living comfortably on the proceeds of the tickets. But we shall soon need more funds, so I would like you to cable me a thousand dollars at once.

Yesterday I heard from the Italian Government. They have accepted my proposition, with only a few minor changes, and I have received the agreement duly signed by the proper authorities. This agreement I am sending along herewith for your ratification. Please don't try to imitate the United States Senate and reject it just to show who is boss.

You will note that we agree to build a factory at Genoa and supply the Italian Government with as many tractors as they want. We agree to pay a minimum wage of one hundred lire a day. For the first year, if we desire, the factory may be a mere assembly plant; parts may be brought in at a very low duty and made into complete machines. We also have free port privileges, which means that we pay no duty at all on parts which are assembled into tractors and then shipped on to other countries. The government agrees to buy eight hundred tractors of various sizes. I am told that if these are satisfactory they will want a good many more.

In addition, I am pleased to report that my Russian venture has begun to bear fruit. Mr. Krimsky appeared the other day and handed me an order for five hundred machines to be used on the new Soviet collective farms. We can save a lot of freight on this order by shipping the parts to Genoa, assembling them there, and shipping the completed tractors on to Odessa. I also enclose an order just received from the Italian Railroad System for fifty tractors—a mere trifle, it is true, but one that we ought not to utterly despise. The Italian orders, of course, are void unless the machines are assembled in Italy.

This morning there was an account in the local newspaper of our proposed factory and the wages we are going to pay. Gadget overheard a group of workmen discussing the news. "They say," said one of them, "that even the man who sweeps out the shop will get one hundred lire a day!"

Truly, history is repeating itself.

Yours,
ALEXANDER BOTTS.

CABLEGRAM

EARTHWORM CITY ILL DEC 19
ALEXANDER BOTTS ROME ITALY

YOUR LETTER RECEIVED STOP CONGRATULATIONS STOP WE HAD NO IDEA YOU COULD GET SUCH LARGE ORDERS AND SUCH FAVORABLE TERMS STOP BOARD OF DIRECTORS AT SPECIAL MEETING HAS DECIDED TO SEND OVER LARGE FORCE OF ENGINEERS AND TECHNICAL MEN WITH SUPPLIES AND MATERIALS TO INSTALL FACTORY AS SOON AS POSSIBLE STOP AM CABLING ONE THOUSAND DOLLARS AS REQUESTED STOP YOU WILL PLAN TO STAY OVER THERE INDEFINITELY

HENDERSON

CABLEGRAM

ROME ITALY DEC 20 1928
GILBERT HENDERSON EARTHTRACT
EARTHWORM CITY ILL

THANKS BUT GADGET AND I HAVE DECIDED TO RETURN TO AMERICA STOP WE ARE LEAVING MARCO IN CHARGE HERE

BOTTS

CABLEGRAM

EARTHWORM CITY ILL DEC 21 1928
ALEXANDER BOTTS ROME ITALY

CANCEL SAILING STOP WE NEED YOU OVER THERE

HENDERSON

RADIOGRAM

S S JUPITER DEC 21 1928
GILBERT HENDERSON EARTHTRACT
EARTHWORM CITY ILL

ALL THE DIFFICULT WORK IS NOW FINISHED SO THE ORDINARY MEN YOU ARE SENDING OVER CAN CARRY ON STOP WE ARE ALREADY ON THE BOAT ONE DAY OUT FROM NAPLES STOP SORRY TO DISOBEY ORDERS AGAIN BUT WE HAVE TO GET HOME SO THAT ALEXANDER BOTTS JUNIOR WILL BE BORN ON AMERICAN SOIL STOP MERRY CHRISTMAS TO ALL

ALEXANDER BOTTS

ABOUT WILLIAM HAZLETT UPSON

William Hazlett Upson was born in Glen Ridge, New Jersey, September 26, 1891. His father was a Wall Street lawyer, his mother a doctor of medicine, and most of the rest of the family doctors, lawyers, college professors, and engineers. To be different, Bill Upson became a farmer—but found the job involved too much hard work. He escaped from the farm by enlisting in the field artillery in World War I.

After the war, he worked from 1919 to 1924 as a service mechanic and troubleshooter for the Caterpillar Tractor Company. "My main job," he said, "was to travel around the country trying to make the tractors do what the salesman had said they would. In this way I learned more about salesmen than they know about themselves." He also became very fond of salesmen. He admitted they are crazy, but maintained they are splendid people with delightful personalities.

In 1923, Bill began writing short stories. In 1927, he created the character Alexander Botts, who has appeared in over a hundred *Saturday Evening Post* stories.

Bill was married to Marjory Alexander Wright. For many years, their home was in Middlebury, Vermont. They had a son, Job Wright Upson, a daughter, Polly (Mrs. Claude A. Brown), and a dog, Shelley. Upson spent his last days in Middlebury with his family, and passed away on February 5, 1975, at the age of 83.

ABOUT ALEXANDER BOTTS*

Requests for biographical material have come from Botts fans, particularly the younger ones who missed many of the earlier stories. From the files of the EARTHWORM CITY IRREGULARS—the national Botts fan club—come the following notes:

ALEXANDER BOTTS was born in Smedleytown, Iowa, on March 15, 1892, the son of a prosperous farmer. He finished high school there, then embarked on a series of jobs—none of them quite worthy of his mettle. In these early days the largest piece of machinery he sold was the Excelsior Peerless Self-Adjusting Automatic Safety Razor Blade Sharpener. He became interested in heavy machinery in 1918 while serving in France as a cook with the motorized field artillery. In March 1920, he was hired as a salesman by the Farmers' Friend Tractor Company, which later became the Earthworm Tractor Company.

On April 12, 1926, he met Miss Mildred Deane, the attractive daughter of an Earthworm dealer in Mercedillo, California. Seven days later they were married. Mildred, later nicknamed Gadget, had attended the language schools at Middlebury College (Vermont) and acted as interpreter for her husband when he was sent to Europe in 1928 to open new tractor outlets there.

Mr. and Mrs. Botts returned from Europe in early 1929 to await the birth of Alexander Botts Jr., who arrived in February along with a twin sister, Little Gadget. Mr. Botts now has been a grandfather for some years.

The adventures of Botts have been appearing in *The Saturday Evening Post* since 1927, over a hundred stories in all. They have been collected into eight books, most of them now out of print. Your local bookstore probably can locate used copies of the books, or, if not, the EARTHWORM CITY IRREGULARS office usually has copies on hand. Write to R. D. Blair, 38 Main Street, Middlebury, Vermont, for information, and—if you're a true-blue Botts fan—ask about joining the IRREGULARS.

**This biography was written by William Hazlett Upson and published in the 1963 book* Original Letters of Alexander Botts *by Vermont Books: Publishers in Middlebury. While the EARTHWORM CITY IRREGULARS are no longer meeting, Upson's local bookstore—The Vermont Book Shop—which opened in 1949 is still open at this address today.*

We are pleased to have presented this collection, which is the second installment of the Alexander Botts stories in their entirety, along with the illustrations that accompanied them in *The Saturday Evening Post*.

We intend to present the full collection of over 100 short stories, including some that were published outside of the *Post*. To see where Bott's next sales call takes him, visit octanepress.com or anywhere books are sold.

Follow Octane Press on Facebook, Instagram, Twitter, or the latest, new-fangled social media platform to learn more!

Made in USA - Kendallville, IN
32508_9781642340396
12 15 2021 1747